C0-AUG-061

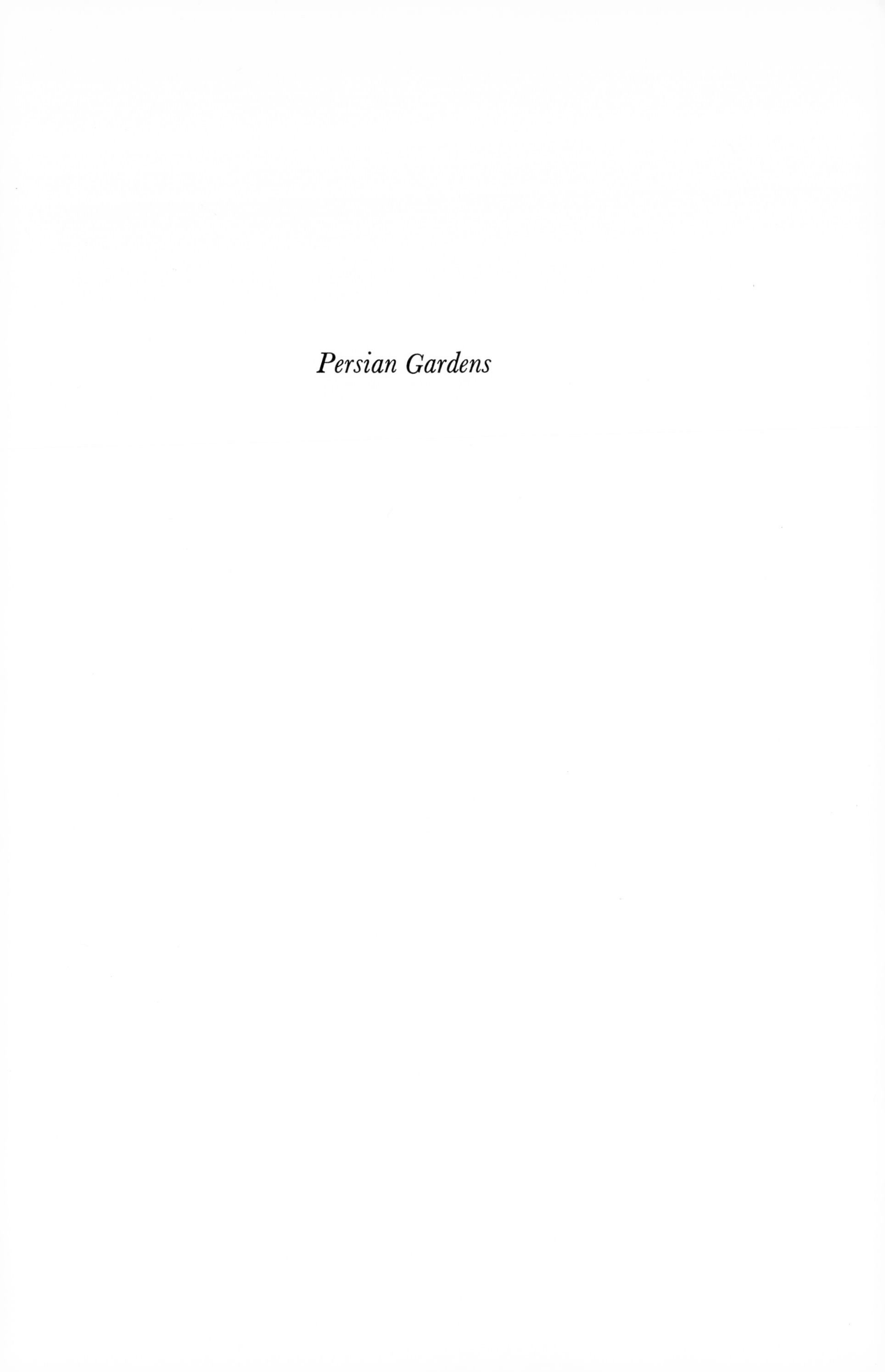

*Persian Gardens*

# PERSIAN GARDENS AND GARDEN PAVILIONS

by DONALD NEWTON WILBER

*Second Edition*

Dumbarton Oaks
Trustees for Harvard University
Washington, District of Columbia
1979

LIBRARY OF CONGRESS CATALOGING IN PUBLICATION DATA

Wilber, Donald Newton.

Persian gardens & garden pavilions.

Bibliography: p. 93.

Includes index.

1. Gardens—Iran—History. 2. Palaces—Iran—History. 3. Pavilions—Iran—History. 4. Garden structures—Iran—History. I. Title.

SB466.I6W5 1978 712'.0955 78-13801

ISBN 0-88402-082-7



Printed at The Stinehour Press

Illustrations by The Meriden Gravure Company

To long-time friends
Ethel and Sherley Morgan
constant enthusiasts of the fine arts
and patrons of generations of students

*Editor's note:*

The text of the present volume is substantially the same as that of the edition of 1962. It has been reissued by Dumbarton Oaks because it has long been out of print and in demand. Minor editorial changes have been made, and the Bibliography has been expanded to cover recent contributions, but no attempt has been made to revise or to update the information in the body of this work to reflect events of the past seventeen years.

*March, 1979*

# Contents

# Color Plates

*Black-and-white plates will be found in a separate section beginning on p. 97*

# Preface to the Second Edition

More than a score of years ago the author made his first trip to Iran. At that time the country was far more remote in space and time than at present when the alarms of international affairs make us familiar with many parts of the world. However, even then I had read of Persian poets and Persian gardens and was prepared to lose myself in the local atmosphere. Nor had I long to wait, for toiling up the long, dusty road from Baghdad to Tehran in the scalding heat of summer I soon learned to seek shelter in gardens along the way, sometimes alongside a teahouse pool banked around with potted plants, and sometimes along a rushing rivulet within a fragrant orchard. Almost at once the identification of the garden with paradise, made by the Persians, seemed natural and appropriate.

After this introduction to the more informal aspects of garden art and design, subsequent trips throughout the country gave opportunities to study gardens of local renown. It was soon apparent that although everyone loved the gardens, no one had done anything to crystallize this feeling. Only a few articles of rather ephemeral nature had been published within Iran or abroad and no accurate plans and drawings of details had been executed. The situation called for action as the finest of the old gardens were disappearing, one by one. As the larger towns of Iran expanded, the gardens on the outskirts were divided up into building lots. In other cases the owners had lost interest in maintaining the pavilions and the grounds, and in many cases, when an older garden would be restored or a new one created, the older Persian features gave place to European design elements. Unfortunately such has been the case at the present royal gardens at Shimeran. In a sense this record of the Persian gardens and garden architecture of the past is intended for the Iranians themselves, in the hope that it may inspire them to preserve and recreate their own heritage.

My personal response to the situation was to make measured drawings of the finest gardens that existed in various parts of the country. However, such drawings, as interesting and unique as they may be, have their more immediate appeal to landscape architects and garden designers so it was necessary also to picture these gardens as they appeared to the persons enjoying them in their finest state. Three sources were available: photographs taken at the present time or as much as a half-century ago; drawings and paintings made a century or more ago; and Persian miniature paintings dating back to the fifteenth century.

The photographs were nearly all taken by the author and are neither so numerous nor so technically perfect as might be desired. As regards the older illustrative material, Persian painters of the nineteenth century are represented in a more comprehensive fashion than in any previous publication, while the reproductions of lithographs and sketches come from the most fascinating of the earlier illustrated accounts of European visitors to Iran. To anyone interested in extending this experience, the author heartily recommends perusal of the monumental publications of Xavier Hommaire de Hell, Sir Robert Ker Porter, E. Flandin and Pascal Coste, Cornelis de Bruyn, and C. Texier. The magnificent plates of these volumes mirror the times, while the works of Chevalier Chardin and Thomas Herbert reflect the intimate details of courtly life within the gardens. Iran has an amazingly rich bibliography and no single work is so dull or so specialized that it does not deserve reading. This comment applies with greatest force to the accounts of those Europeans who came to Persia as if it were the end of the world, and remained to make careful records of marvels which surpassed those of their native London, Paris, or Moscow.

The illustrative material includes Persian miniatures, works of art at once precise, delicate, detailed, and faithful to life. Their painters depicted kings, princes, heroes, warriors, and even saints and angels, enjoying themselves in gardens; the miniatures are marvelously successful in catching the atmosphere of repose, of gaiety, and of delight. They portray the details of fountains and of pools, of tents and of kiosks, of pavilions and of palaces, of composed vegetation and of exuberant nature; but they show the momentary aspect of the subject, the garden as it is enjoyed by particular persons at a moment in time. What these miniatures do not show is exactly what we wish they portrayed: perspectives of gardens and pavilions displaying the general plan and the relationship of details to the larger composition. It would seem as if these painters might have given some thought to the future, especially since their counterparts in India, whose style derived from Iran, did execute aerial views of gardens, lakes, and palaces, and of gardeners at work in their domains. That these painters themselves loved the gardens seems reflected in the manner in which flowers, flowering shrubs, and flowering trees crowd their pictures. In fact the portrayal of flowers developed into a separate minor art during the eighteenth and nineteenth centuries and these flower paintings have become, as reflected by their current scarcity, the most popular articles ever produced in Iran.

The plans and illustrations within these covers are not intended to portray something exotic and unrelated to our climates and interests. All gardens are made up from a few basic features—open spaces, massed foliage, flower displays, water, and vistas—and the gardeners who read this book should come across many details and ideas that can be incorporated into their own kinds of gardens. From the broader point of view, the Persian concept of garden and pavilion should find sympathetic appreciation from all interested in today's trend toward out-of-doors living. As contemporary architects design houses in which pools and plants invade the living quarters, they are in tune with the Persian designers whose garden pavilions drew no firm line between the interior and the garden itself. Persians enjoy nature in their gardens in an easy and casual manner and never romanticize the settings. In this respect their attitude is much closer to our own than would have been the case a half-century ago in Europe and the United States.

Thanks are due to individuals and institutions. Warmest thanks go to a friend of long standing, Muhammad Taqi Mustafawi, formerly Director of Antiquities in Iran, who supplied photographs of material in official collections, collected photographs of paintings by Persian artists, and granted permission for reproduction. Appreciation is also extended to the Metropolitan Museum of Art, the Freer Gallery, the Walters Art Gallery, the British Library, the Victoria and Albert Museum, and the Library of Congress for permission to publish photographs of miniatures and other objects in their collections; as well as to Joseph M. Upton for his advice on dealing with the gardens at Shiraz.

The author is most grateful to the Center for Studies in Landscape Architecture, Dumbarton Oaks, for its decision to reissue this work in a modified form and, in particular, to Professor Elisabeth B. MacDougall who undertook this challenging task, and to Lois Fern for her editorial assistance.

As noted above, nearly all the plans and details are previously unpublished. Field notes and measurements were taken by the author and Margaret Surre Wilber and the drawings themselves made by the same pair. Some of the material was collected on expeditions made to Iran by the Iranian Institute and led by Arthur Upham Pope, for whose cooperation the author also expresses gratitude.

Sources for the illustrations are indicated in the accompanying captions, with the exception of those either by the author or coming from his collection. The following abbreviations have been used in the captions:

ASI: Archaeological Service of Iran

Coste: Coste, Pascal. *Monuments modernes de la Perse, mesurés, dessinés, et décrits*. Paris, 1867

Flandin: Flandin, Eugène, and Coste, Pascal. *Voyage en Perse: Perse moderne*. Paris, 1854

Hommaire: Hommaire de Hell, Xavier. *Voyage en Turquie et en Perse*. Paris, 1854–60

Kaempfer: Kaempfer, Engelbert. *Amoenitatum exoticarum politico-physico-medicarum, fasciculi V, quibus continentur variae relationes, observationes et descriptiones rerum Persicarum et ulterioris Asiae*. Lemgo, 1712

The terms Persia and Iran, or Persian and Iranian, are used interchangeably throughout the text. Readers will know that the country has always been called Iran by the people themselves, while the word Persia derives from the single province of Fars, or Pars, of which Shiraz is the center.

DONALD N. WILBER
June 1978

# Persian Gardens and Garden Pavilions

CHAPTER ONE

# Persian Gardens and Paradise

Throughout the Islamic centuries, from the Arab conquest of Iran until the present day, gardens have represented images of paradise to the Persians. Beyond question it was the fact that the plateau has always been relatively arid and treeless that gave the gardens such a supreme value. The origin of the life-giving aspect of the gardens may go back to the most remote periods. As early as 4000 B.C., when hunters were moving down from the mountains and settling as agriculturists in the long valleys of the lofty Iranian plateau, they depicted aspects of their lives and beliefs upon the earliest decorated pottery. Some of these bowls show pools of water overhung by the tree of life. Others show the world as if divided into four quarters, and some of these patterns have a pool at the center. This type of cross plan, in which one axis may be longer than the other, was to become the standard plan of the Persian garden under the name *chahar bagh*, or four gardens. The plan form itself was crystallized at least as early as the Sassanian period (A.D. 224–642) in the hunting park which had a palace or pavilion at the intersection of the axes.

The high regard, even worship, of the flowing pool that existed before the dawn of history did not wane in later times. In the Sassanian period the rulers favored one special kind of site for the great rock carvings depicting their triumphs and their religious devotion. Each such site was a spring-fed pool at the base of a sheer face of rock. A thousand years and more later a Qajar ruler of the nineteenth century selected exactly such a site, at Rayy just south of Tehran, for a carving showing himself on his throne.

Since water is the key to all Persian garden development, the source of this water is vital. Over most of the plateau, rainfall is less than a dozen inches a year and the topography and climate recall those of Arizona and New Mexico. Only along the Caspian Sea is there a heavy rainfall and a semitropical climate. On the plateau proper the principal source of water is the snow which falls on the high mountain peaks during the winter months.

In the spring the melting snow creates thousands of mountain streams which dash down rocky valleys. Few of these streams are perennial and few become respectable rivers winding their way to the seas. Most of the streams and rivers flow into the interior deserts of the country where they are absorbed by the hot sand and gravel. During a few months of the year when water is available in these streams, channels are led off to irrigate fields and gardens. Springs and wells may also supply water but the principal source comes from a typical Persian device called the *qanat*, or *kariz*. The method itself appears to date back to Achaemenid times (sixth century B.C.).

To construct a qanat line, a spot is chosen at the base of a snow-capped mountain some distance from the place of habitation and at a higher level on the valley floor. A master shaft is sunk to the permanent water table, a distance which may be hundreds of feet, in porous rock or gravel. The true elevation of the water table and a point where the water is to be delivered on the surface must be in such a relation that a tunnel with a very slight slope will join them. This tunnel is painstakingly dug by hand from its point of emergence toward the base of the master shaft. At intervals of every fifty yards or so, a shaft will be dug from the ground to the tunnel in order to provide a means for bringing up the excavated material and for supplying air to the diggers. The length of the qanat line may vary from a few hundred yards to several miles. Once the line has been completed it will supply clear, cold water in about the same volume the year round; a good volume is four cubic feet a second, sufficient for the periodic irrigation of two hundred acres of farm land.

The tunnel of the qanat becomes an open channel a few hundred yards from a village. On this upper slope the abundant supply is used first by the owners of the qanat or by those buying the largest quantity of water: the channel flows into great pools (Plate I) in front of pavilions which are surrounded by orchards and gardens, and then the water runs on to turn the wheels of flour mills. On the outskirts of the more densely populated area it divides into a number of rivulets which border narrow lanes. Below the village the many small channels spread out across the cultivated fields until every precious drop has soaked into the soil.

The distribution of the water supply of a village, or even an urban area, is in the hands of the *mir ab*, or master of the water. His authority must be respected by all, since the most frequent and serious of all quarrels in Persian villages are over the distribution of water. The owners of houses, gardens, and fields rent the flow of a channel for a certain length of time each day or each week. The allotted time is measured in various ways: one common device is to check the time required for a round brass bowl with a pinpoint hole through the bottom to sink in a larger water-filled bowl.

Water being available, a garden is inevitable, and it seems reasonable to begin here with a consideration of the basic terms descriptive of the garden and its features.

The terms which the Persians use to describe their gardens and the flowers which grow in them reveal their absorbing interest in the subject. While much of the language was impregnated with Arabic root words centuries ago, most of the terms relating to flowers and floral growth are from pure Persian roots.

The garden is frequently called *bustan*, or *bostan*, and these words were taken over as being gracefully descriptive of collections of poems. It must be admitted that one of the common words for a garden, *bagh*, is Arabic. It is used with Persian endings to form a number of terms, such as *bagher* which denotes an orchard or garden laid out in compartments. Also, the gardener himself is the *baghban*.

In the realm of Persian flowers the rose predominates, and it seems natural that the Persian word *gul* should be both the word for rose and the general term for flower. The many varieties of roses have their own double names: *gul-i-surkh* for the red rose, and *gul-i-zard* for the yellow one. Nearly every individual flower includes "gul" in its name, for example, the *gul-i-narges*, or narcissus. The names of some flowers, such as the tulip, may be used with or without "gul"—*gul-i-laleh* or *laleh*—while a garden full of tulips is called a *lalehzar*. Naturally enough, many of the names of flowers are descriptive of their appearance, and one interesting sidelight is that certain flowers remind the Persians of things quite different from what we see. For example, the dahlia is the *gul-i-kukab*, or star flower. The pansy is called the European violet and the periwinkle is the *gul-i-telgrafi*, or telegraph flower. Why?

"Gul" appears in a number of other compounds: *gulistan* is the rose garden, while *gulshan* and *gulzar* describe flower gardens. A *gulgasht* is a pleasure ground; a *gulkari*, a flower bed; and a *guldasta*, a nosegay. Even the song of the nightingale is the *gulbang*, or flower cry.

Many a Persian could call himself a *gulbaz*, flower lover or rosarian. The perfume of the rose was not confined to the garden. Quantities of rose water, or *gulab*, were pressed from the double pink rose (*rosa centifolia*), and Kashan was noted, and still is noted, for this product. The rose water was kept in distinctive spouted vessels of blown glass and it was the custom to sprinkle guests with rose water, but in such a dilution that clothes were not spotted. Rose preserves were much favored; one kind might be sweetened with honey and another with sugar; concentrated rose water still flavors a variety of desserts, including sherbets and pastries.

Naturally enough the rose plays an important role in poetry in such figures as metonyms, similes, and metaphors. The beloved's face, her cheeks, her forehead, or her entire figure are identified with the rose, her tears with rose water, and the world wherein she dwells with a rose garden.

What of the flowers themselves? Since the French jeweler Chardin, who visited Iran in the seventeenth century and was such a keen observer of the local scene, has left us a chapter on Persian gardens, a summary of his words offers an excellent introduction.

"There are all the kinds of flowers in Persia that one finds in France and Europe. Fewer kinds grow in the hotter southern parts, but by the brightness of coloring the Persian flowers are generally more beautiful than those of Europe. Along the Caspian coast there are whole forests of orange trees, single and double jasmine, all European flowers, and other species besides. At the eastern end of the coast, the entire land is covered with flowers. On the western side of the plateau are found tulips, anemones, ranunculi of the finest red, and imperial crowns. Around Isfahan jonquils increase by themselves and there are flowers blooming all winter long. In season there are seven or eight different sorts of narcissus, the lily of the valley, the lily, violets of all hues, pinks, and Spanish jasmine of a beauty and perfume surpassing anything found in Europe. There are beautiful marsh mallows, and, at Isfahan, charming short-stemmed tulips. During the winter there are white and blue hyacinths, lilies of the valley, dainty tulips, and myrrh. In spring yellow and red stock and amber seed of all colors, and a most beautiful and unusual flower called the clove pink, each plant bearing some thirty flowers, blooms.

"The rose is found in five colors: white, yellow, red, Spanish rose, and poppy red. Also there are 'two-faced' roses which are red on one side and yellow on the other. Certain rosebushes bear yellow, yellow-white, and yellow-red roses on the same plant."

Chardin's pleasant catalogue can easily be expanded. From the pages of an early English traveler may be added the clove gillyflower, cornflower, hollyhock, marigold, poppy, primrose, saffron, lilac, and daffodil.

From other observations may be added the grape hyacinth, the miniature Persian iris, species tulips, the carnation, pink, and evening primrose. And, in addition, one finds the primula, oenothera, delphinium, tuberose, and musk flower. However, with this list of some thirty-five kinds of common flowers one may stop and not resort to copying from the Persian dictionary the names of some sixty others which are far from rare.

A subject on which little reliable information is available is the relationship of Persian flowers to their habitat—which flowers are native to the country or region and which were imported in more recent times. In certain cases the Persian name of the flower reflects the need to make a descriptive name for a plant not known before. The Persian names of varieties of roses give some indication of how complicated can be the question of the origin of names and hence of species. For example, the hybrid tea rose is known as the Bombay rose or egg rose, both terms suggesting that it was in imported form. However, the tea rose itself is also called the "tea" rose in Persian—*chai* from the Chinese, while the China rose has that same descriptive name. There seems to be agreement that the Persian double yellow rose (*rosa hemispherica*) is a native variety which was taken to Spain hundreds of years ago and to other parts of Europe within the last three centuries.

It seems difficult to keep from saying just a little more about the Persian rose. Mention is made of the hundred-petaled rose as early as the eleventh century, during the reign of Mahmud of Ghazni, and the same species was recorded at Isfahan in the seventeenth century. Perhaps a species like this was grown in the plantations where several thousand pounds of roses were gathered each day for the preparation of attar of roses. The rose cast its spell upon other flowers. The emperor Babur related his pleasure in the discovery of a rose-scented tulip.

Gardeners are mentioned rather infrequently in Persian history and literature, and it is not certain whether the local Zoroastrians, or *gabris* as they are commonly known, were renowned as gardeners in all the centuries after the Sassanian empire when Zoroastrianism had given way to Islam. Chardin does say that the village of Nejafabad, near Isfahan, was the habitat of gabris, who were known as cultivators. However, throughout the nineteenth century the gabris were famed as gardeners and also as the most skilled diggers of qanat systems.

The routine of the gardener in Iran is much the same as anywhere else: he tends flats and potted plants in greenhouses, prunes and espaliers fruit trees, trims hedges, and sweeps walks. At work with the dawn, he is not in any hurry but takes a genuine delight and satisfaction in his work. Although a garden may be noted for some special feature, such as spreading wisteria or beds of giant pansies, the Persians never devoted serious attention to improving species or to hybridization. This may have been because they admired nature in its most familiar and luxuriant aspects. There may be something of a parallel here with the Persian attitude to fine arts, for in architecture and decoration favorite motifs continued to be used almost unchanged over long periods of time.

Because the Persians were so fond of gardens and flowers, they took steps to recreate a floral atmosphere during inclement seasons. The manner in which artificial flowers and trees were made and used is a subject in itself, and only the most sketchy mention of it can be made in these pages. One Persian manuscript, in the India Office Library at London, gives directions for making artificial flowers.

From early times there was an intimate relationship between the ruler, his throne, and an artificial tree. The poet Firdawsi describes a tree made for the king Kay Khusraw which had a silver trunk and gold and ruby branches, the branches bearing hollow fruits filled with a mixture of musk and wine. In the Ghaznavid period (the eleventh and twelfth centuries) the court was enhanced with trees of gold flanked by artificial narcissi in pots of silver. At the Mongol court in Iran, during the thirteenth and fourteenth centuries, similar trees were featured, and the most elaborate one of all appears to have been made for Timur and set up in the royal tent at Samarqand. The height of a man, this man-made tree bore leaves like those of the oak, while its fruits were of pearls, rubies, emeralds, sapphires, and turquoises. Little birds of colorful enamel perched upon the branches.

The trees and the flowers, the fruit and the miniature gardens were long produced by distinctive craftsmen called the *nakhlband*. This term has an Arabic root and a Persian ending meaning "a maker of artificial flowers," but it may also have the meaning of "a festoon maker." Paper, paste, wax, and paint were the ingredients of this minor art, with the most costly and elegant productions carried out in colored wax. However, by the time of the Qajar period (1779–1925), the avenue of approach to the ruler on his outdoor throne was lined with vases containing artificial flowers of paper. And such paper flowers are still popular in Iran.

The larger the garden, the more certain it was to have an extensive tract of orchard with the fruit trees spaced at regular intervals in regular rows. Such orchards supplied the owners of the garden and provided a cash crop. The fruits included apricots, plums, prunes, cherries, quince, peaches, mulberries, pears, apples, figs, oranges, sweet and sour lemons, and limes. The apricots of Iran are especially juicy and flavorful and are dried in enormous quantity. Apples are probably the least satisfying in a harvest of fruit that is noteworthy for its fine flavor. The citrus fruits are grown either along the Caspian coast or in the south and southeastern areas of the country. Splendid pomegranates grow on vigorous bushes.

Grapes in at least a dozen different varieties were common in gardens throughout the land. As detailed elsewhere, Shiraz was famous for its wine, but Qazvin, Tabriz, and other towns produced their own varieties as did Isfahan, where small, white, seedless grapes were pressed. Bunches of grapes left on the vines were protected by cloth bags and could be picked until well into the winter; after that they were stored in cool, dry rooms until spring, as were melons that had been picked in the fall.

Palm trees appear from the Persian Gulf up into the central part of the plateau, but the date palms bear satisfactorily only in the more southern regions. Among the nut-bearing trees are the English walnut and the hazelnut, but the species most common and most popular are the pistachio and the almond. Just as the first fruits of the orchard are so prized, so the Persians pick the almonds before they are ripe, and, after removing the soft green husk, eat tender shell and nut together.

All the common trees of Iran are the familiar ones: the plane, poplar, elm, willow, maple, sycamore, ash, cypress, and pine. Of these the plane tree, the poplar, and the cypress are the most admired and desired. The giant plane trees cast wide areas of shade and enjoy as well the reputation of protecting their place of habitat against fevers and other air-borne infections. The poplars grow with amazing rapidity and are valued for purposes of construction, while the cypress is the tree most cherished by poets.

Birds and animals were kept in the larger gardens, and at Isfahan in the time of Shah Abbas (ruled 1588–1629), two of the most prominent positions along the avenue of the Chahar Bagh were taken up by an aviary and a zoo. Gazelle were (and still are) most frequently seen in gardens, captured young and kept until increasing intractability destined them to other purposes.

Swans, pigeons, pelicans, and ducks swam upon the pools, but one bird was so extravagantly admired that it might be named Iran's national bird. This is the *bulbul*, or nightingale. The finest gardens had the densest shade to attract the most nightingales, and for the Persians there is no finer pleasure than to wander through a moonlit garden where the nightingales are in full voice. There is even a word for a garden so favored: a *bulbulistan*, or place of nightingales.

Just as there was always a vogue for the creation of artificial plants and flowers to be enjoyed when the gardens were not in full bloom, so artists were encouraged to record all the varieties of flowers. At early dates in the Islamic period careful drawings of plants and flowers appear in the works dealing with medicine and with the medicinal properties of plants. In later periods the plants shown became less stiff and stylized and a version of Dioscorides' *Materia Medica*, copied and illuminated in the time of Shah Abbas, displays charming pictures of growing flowers.

It may have been the early nineteenth century before artists concentrated a great deal of attention upon producing very naturalistic paintings of flowers. Precise information on this subject is lacking, because writers and collectors devoted little attention to the flower paintings until they disappeared from the market. Museums and private collectors paid large sums for paintings of the Timurid period (the late-fourteenth and fifteenth centuries) and the Safavid period (1500–1736) and ignored the genre art of the Qajar period. This later work was acquired by hundreds of travelers and tourists, so that today flower paintings are very difficult to find in Iran, while they are rarely found in museums anywhere.

From the limited information available, it would appear that the painting of flowers was part of the training of all apprentice artists. The high point of this art was the execution of albums of flower paintings. The albums might contain a miscellaneous collection or might concentrate on a specific subject—such as the album which illustrated the twenty-nine floral varieties grown in the garden of Sa'adatabad at Isfahan. It is certain that thousands of single sheets depicting flowers were turned out in the period from about 1800 until 1875: some were done in India-ink wash, others in a single color, but most were vivid and gay in many colors—bright enough to cheer an overcast winter's day.

Flowers escaped from paper and appeared on book covers, pen cases, mirrors, and numerous other articles which were made of lacquered papier-mâché. This was a minor art whose products were at the same time most reflective of the period and utilitarian. Certain towns turned out their own painters with distinctive styles: Shiraz was a center of flower painting and an entire series of floral-painted book covers has survived. These covers enclosed manuscript copies of the Qoran, and examples collected at Tehran are dated from 1757 to 1858. Some lack an artist's signature, but others reflect the careful style of a recognized artist (Plate 2). By far the most popular of all these cover artists was Luft 'Ali Suratgar, who was active at the middle of the nineteenth century.

While it is well known that European artists and examples of their work influenced the murals of the Safavid period at Isfahan, Ashraf, and other sites, the Persian artists did cling to local techniques and styles of flower painting. It may, therefore, come as a surprise to read the account of Cornelis de Bruyn who visited Isfahan about 1735. He called on a local artist who was engaged in copying for the Persian ruler the illustrations in a book on flowers printed in Europe. During the Safavid period a number of species of flowers unknown to Iran were introduced from Europe and it may well be that the Shah and his court placed their orders from such hand-painted catalogues.

The gay, beautifully accurate representations of flowers on woven textiles of the seventeenth and eighteenth centuries, often enriched with metallic threads of silver and gold, have survived in considerable number—too many to describe or picture. However, for another type of artistic production a single piece may serve to illustrate details common to many other examples. This is the case for the so-called Bokhara embroideries, and the illustration (Plate 3) shows a splendid example woven early in the nineteenth century and now on display in the Metropolitan Museum of Art. On a piece of cotton an entire catalogue of flowers is displayed with painstaking accuracy and affection. Blue, red, and green dominate the palette and scarcely any of the flowers are shown more than a single time: a botanist might identify at least twenty-five different flowers. These Bokhara embroideries display a steady development from a prototype form: the earliest ones have the most naturalistic flowers, while by the end of the last century the floral motifs and elements had become very stylized.

In contrast to the Bokhara embroideries, the later flower paintings display just the opposite tendency. The later pieces show more realistic, more full-blown flowers—some beyond their prime and with petals about to fall—as in two examples (Plates 4, 5).

Gardens and garden elements flourished independently of those cultivated on the ground itself. Gardens influenced art forms and were in turn influenced by certain artistic motives and styles.

Carpets and gardens display a charming affinity. As early as the Sassanian period great hangings portrayed the flowering spring, and textiles of all later periods featured floral elements. But it was not until the early seventeenth century and the reign of Shah Abbas that the garden carpet came into its own. These carpets were called chahar bagh (four gardens) carpets because they reflected the typical division of the garden into four sectors related to the major and secondary axes. One of the very earliest and certainly the most magnificent of these carpets is now in the Jaipur Museum in India. Made prior to A.D. 1632, it was recorded that year in the Amber Palace of the maharajas of Jaipur as a foreign production, and there is good reason to believe it was woven in the vicinity of Isfahan and that its design was inspired directly by the contemporary gardens of Shah Abbas. The carpet, which is twenty-eight feet in length, represents a very large garden, so extensive that each of its four sectors was subdivided by water channels into a number of other compartments. Fish swim in the channels, birds nest and fly in the heavy foliage, and a great variety of trees and flowers crowd the compartments. The huge central pool is overlaid with the image of a garden pavilion seen in direct elevation. Probably this pavilion stood behind the pool. From the representation it is apparent that the pavilion was square in plan, open on each side, decorated with murals reflecting the spirit of the garden, and crowned with a delicate blue-tiled dome.

A considerable number of garden carpets survive in private and public collections: few are of the early seventeenth century and most of them are assigned to the eighteenth and early nineteenth centuries. Fairly characteristic of these later examples, which are frequently more stylized in design, is a carpet believed to have been woven in the mid-to-late eighteenth century (Plate 6).

The so-called vase carpets and the hunting carpets of Iran do not repeat the garden pattern but they do display a lively interest in foliage and flowers. The renowned hunting carpets show personages in flowering landscapes, and occasionally a garden pavilion appears amidst the foliage. Of course, the great majority of the Persian carpets of all periods feature floral elements in a rich variety of patterns and combinations. These designs vary from closely compartmented ones to free-flowing forms. However, nearly all patterns display a central motif and axial symmetry so that the floral elements fit into a formal framework. It is this usual carpet pattern which is reflected in garden design; beds of variegated colors and forms, and borders of small shrubs repeat the design of the carpet. It is doubtful whether this manner of copying rug patterns was in use as early as the Safavid period. None of the earlier travelers appears to have noticed floral beds of this type and it is possible that the idea did not take hold until the nineteenth century. Today many of the formal gardens display these "carpet" beds of flowers, and for the taste of some they seem rather too bright and garish.

Since Persian gardens make so much of water, it may come as a surprise to find gardens in which no water was used. These are very special cases and appear to be confined to the province of Khorasan, in northeastern Iran. At the town of Fariman water is so scarce that the gardens are created in colored stones collected from the dry stream beds. The pattern of one of these gardens in miniature contains motifs from carpets but is not a precise copy of a carpet design.

Pools are of paramount importance in Persian gardens and, in fact, in every Persian household. They range in size from the *daryacheh*, or little sea, to the smallest circular or rectangular pool in the courtyard of the humblest village house. Where these pools are a feature in the courtyards of village and town houses, they are replenished at fixed intervals from open channels flowing down the streets and lanes. Contrary to the belief of many foreigners, the water in these pools is not used for drinking but constitutes a reservoir for watering the garden and laying the dust on the paths and terraces.

Probably the word most frequently used for the pool is *hauz*, an Arabic word which appears in Persianized forms as the *hauzcheh*, or little pool, and *hauzkhaneh*, or pool within a house. The Persian word for pool, *istakhr*, has also the connotation of a lake. The Persians may refer jokingly to a tiny pool as a *darya*, or sea. The indication that playing fountains and jets of water were developed outside of Iran comes from the use of an Arabic work *favara* for these features, together with the apparent lack of a corresponding Persian word.

The pools themselves are to be found in a great variety of sizes and shapes; most frequently they are situated on an important axis of the courtyard or garden but with their own long axis at right angles to that of the larger area. Always they are on a slight slope, so that they are gravity fed and the overflow distributed by gravity, often with short falls or chutes of water below the pools. Typically Persian is the manner in which the sides of the pools are built up above the surrounding level, with an exterior gutter to carry off the overflow. Given an adequate water supply, these pools are always brimming over so that from any little distance one sees a sheet of water which appears completely unconfined.

Sources for the shapes of these pools come from existing gardens, from illustrations showing vanished pools and gardens, and from the miniature paintings dating back to the fifteenth century. So-called free forms do not appear in Persia, where forms based upon the rectangle and the circle predominate. Pools of the present or those known to have existed in important gardens illustrate how interest and variety were achieved (Plate 7). Characteristic of the Safavid period at Isfahan was the use of tiered pools in as many as four tiers, with each shape at the higher levels of a form different from the ones below. The catalogue of pools may be completed from an examination of those Persian miniatures in which pools are shown. Most popular was the lobed or scalloped shape. A manuscript of the mid-fifteenth century shows a pool with twelve lobes, and paintings of later periods illustrate eight-, nine-, and sixteen-lobed pools. A fine miniature executed at the end of the fifteenth century has a pool edged with eight trilobes.

Jets of water were featured in outdoor and indoor pools. The great Safavid garden of the Hazar Jarib at Isfahan had as many as five hundred jets of water. Such jets were fed by lead pipes and in the more elaborate groupings of nozzles, combinations of patterns could be created by opening and closing valves to each pipe. We hear of such a system from an account of a visitor to the Bagh-i-Takht, or Garden of the Throne, at Shiraz. His host was anxious to display patterns of spouting water and accomplished this effect by means of servants who dashed from nozzle to nozzle, holding wads of cloth over those openings not required at a given moment: this haphazard method was used after the central control system had fallen into disrepair.

Upon jets of considerable force red apples defied gravity, while other jets sprang from the mouths of lions. At Isfahan, massive column bases were laboriously pierced with holes so that water shot directly from these columns into the pools on which they bordered. It is apparent that a considerable use was made of water under pressure. At the Ali Qapu Palace at Isfahan a fountain played in a pool on the open porch on the fourth floor, and a contemporary account states that the water was raised by oxen power. It is probable that an endless chain of buckets plunged into a cistern below ground level and then raised the water to a tank on the fifth or sixth floor of the palace.

In general, the sheen of water in the larger pools was admired for its dark, reflecting quality. The open channels were lined with blue tiles, but the larger pools were not intended to be crystal clear—rather, to suggest dark, unfathomable depths—and they were not cleaned out too frequently. Upon occasion the surface of the water was treated to obtain special effects. Rose petals might be strewn across the surface or, in the evening, candles set adrift on tiny rafts. Early in the nineteenth century a European traveler saw a pool, the surface of which had been divided by switches or rods into a number of compartments (Plate 8). In each compartment floated different flowers. This effect was probably inspired by the interest in imitating carpet patterns and colors in other media.

The sound of water was always enjoyed. The practice of building gardens on slight slopes gave rise to rushing, gurgling, splashing water channels, while changes of level might be marked by a stone-lined incline decorated with a scale pattern, so that the smooth flow of the rushing water was broken up into a flashing design.

The image of paradise long ago established in the minds of the Persians is that of the most delightful gardens. Paradise is *bihesht* in Persian, and this word appears in the names of many of the gardens. Another Persian word, *firdaws* has the meanings of both garden and paradise.

The poets delighted in making comparisons between earthly gardens and the heavenly paradise promised by the Qoran. As the supreme model the poets took the garden of Eram, or Iram, which is mentioned in the Qoran as "Iram adorned with pillars." This garden was built in the Yemen by Shaddad, king of the Adites, to rival paradise, but no sooner was it completed than the ruler dropped dead and the garden vanished. Whole manuscripts were composed in praise of that fabulous garden and many later ones gained reflected glory by taking its name. Such a garden at Tehran was said to exceed the original in splendor.

For hundreds of years the Persian poets praised gardens and the beauty of nature. The gardens were looked at from several points of view: as oases of quiet and retreat, as settings in which flowers, water, and trees could be lovingly described, and as examples of royal pomp and magnificence. In this vast outpouring of praise the Persians established close communication with nature long before this type of relationship between man and his environment became common to many other parts of the world.

On another page, mention has been made of the fact that collections of poetry were entitled bustan (orchard), and gulistan (rose garden). In the *Gulistan* of Sa'di, poet of Shiraz, there is a charming passage which expresses a deep, personal delight in the arrival of spring: "God has told the chamberlain of the wafting breeze to spread an emerald-green carpet and has ordered the nourisher of the spring clouds to bring to fruition the daughters of the plants in the cradle of the earth. God has had the trees put on their New Year's robes, a green-leaved clothing, and has put on the heads of the children of the branches the hat of blossoms to welcome in the springtime. By God's will the juice of the sugar cane has taken on the sweetness of honey, and through God's encouragement the seed of the date has become a tall palm tree."

Firdawsi, the great epic poet of Iran who wrote his *Shah nama*, or History of the Kings, about the year A.D. 1000, delighted in describing the beauties of nature. Writing of an area along the Caspian Sea, he says:

Mazanderan is the bower of spring . . .<br>
Tulips and hyacinths abound<br>
On every lawn; and all around<br>
Blooms like a garden in its prime,<br>
Fostered by that delicious clime.

The nightingale sits on every spray
And pours his soft melodious lay;
Each rural spot its sweets discloses;
Each streamlet is the dew of roses. . . .

In another place the garden of the daughter of the ruler Afrasiyab—a garden of perpetual spring—is described:

It is a spot beyond imagination
Delightful to the heart, where roses bloom,
And sparkling fountains murmur—where the earth
Is rich with many-colored flowers; and musk
Floats on the gentle breezes, hyacinths
And lilies add their perfume—golden fruits
Weigh down the branches of the lofty trees. . . .

Another of the renowned poets of Shiraz, Hafiz, finds communion with nature more personal and intimate in such lines as these:

Oh! bring thy couch where countless roses
The garden's gay retreat discloses;
There in the shade of waving boughs recline
Breathing rich odors, quaffng ruby wine!

In another ode Hafiz says:

When now the rose upon the meadow from Nothing into
Being springs
When at her feet the humble violet with her head low in
worship clings . . .
Earth rivals the Immortal Garden during the rose and lily's
reign;
But what avails when the immortal is sought for on this
earth in vain?

This same mood of searching and longing comes to a sharper, more bitter focus in the verses of Omar Khayyam:

Iram indeed is gone with all his Rose,
And Jemsyd's Sev'n-ring'd Cup where no one knows;
But still a Ruby kindles in the Vine,
And many a Garden by the Water blows.

But in spite of man's departure from the scene the rose remains:

> Look to the blowing Rose about us—'Lo,
> Laughing' she says, 'into the world I blow;
> At once the silken tassel of my Purse
> Tear, and its Treasure on the Garden throw.'

In the nineteenth century much of the poetic writing displayed an interest in unusual verse forms, exaggerated metaphors and other devices, but a few writers continued to speak directly and clearly. Such a one was Qa'ani who was very active about the middle of the century and who penned these words:

> Are these violets growing on the ground on the brink of the streams,
> Or have the houris (of Paradise) plucked strands from their tresses?
> If thou hast not seen how the sparks leap from the rock,
> Look at the petals of the red anemones in their beds
> Which leap forth like sparks from the crags of the mountains!

Quite in contrast is the output of the poet laureate of Fath Ali Shah, a Fath Ali Khan from Kashan who wrote under the pen name of Saba. One of his poems was a 150-line panegyric in praise of a garden made at Tehran for Fath Ali Shah and named after the legendary garden of Eram mentioned earlier. There appears to be no other record of this ephemeral garden. Quoting about a third of this poem, and including notes of explanation, should serve to illustrate what was admired in a garden at this period and to what lengths of hyperbole a poet felt he must go in order to please the ruler.

> When the ruler decided to make flourishing the region of Rayy,[1]
> There were to be lofty castles to raise their summits to the castles of the sky.
> The king established the plan of this garden of his kingdom,
> So that the trees should be as green as the fortune of the king.
> This garden Eram is comparable to the old garden Eram:

[1] South of Tehran; site of an ancient capital of Iran.

However, the ancient garden is grieved at comparing itself
with the new,
In regard to those trees which have opened their green
umbrellas to the Heavens,
And through which the sunshine makes a complicated pattern
Like the arrangement of the stars in the evening sky,
Or like the scattering of coins by the royal hand.
If a spray of the flowering bushes of this garden is not like
Moses' bright hand,
Then why is its flowery aspect like the dawning of the stars,
And if the wind has not swept through the enclosure like
the breath of Jesus,
Then why does the breath of its breezes bring life to the
lifeless?
The famed Tuba tree[2] is like a bramble bush beside these
trees,
And the fountain of life like poison compared to its delicious
waters.
The Tuba and the Sedra[3] trees admire the trees of this
garden;
Kasr and Tasmin[4] envy the spray of these fountains.
If this garden is not Paradise, then why like Paradise,
Does it create happiness through its inner life?
Its enclosure is like the precincts of a drinking club,
And the young trees, like the imbibers, are full of gaiety;
And if the trees of this garden are not intoxicated, then why
like drunkards
In conviviality do they embrace each others' necks?
A building is built by royal order in that garden
Whose glittering pinnacles make the sun seem dark;
Although the king of the seven regions has called it the
"Eight Paradises,"[5]
A name to be inscribed on the surface of Heaven,
Still its height is loftier than the seven heavens,

[2] A tree which stands in paradise with angels sitting on its many branches.
[3] A tree situated in the fourth of the Muslim heavens.
[4] Springs or lakes in paradise.
[5] In Persian *hasht bihesht*, a frequently popular name for a garden.

And its extent is greater than the original eight paradises.
No wonder then that when Mani[6] and Azar[7] saw its beautiful paintings,
They broke their brushes from shame.
Also in this garden was a flowing fountain like Tasmin
That has aroused the envy of the clear well of Zemzem.[8]
The clear water of that pool is like the life-giving breath of Jesus,
And it seems that Mary may have washed her virgin-pure clothes in it.
This playing fountain, like the hand of the Shah,
Fills the earth and the sky with a shower of pearls.
I asked, "Is this the life-giving water?" and the answer came, "Yes."
I asked, "Is this garden Paradise?" and wisdom replied, "Certainly."
The king of kings called this garden Eram,
Although its namesake was not as wonderful as this new one.
When it was completed and its date was to be recorded,
All the poets took thought as to how it might be expressed;
Wisdom showed me the best solution and said, "Oh! Saba,
Tell the king of the world, the honor of nations,
That the breeze[9] carried along the flowers through the garden Eram,"
And said for the date, "Let the garden Eram remain flourishing."[10]

These passages from the Persian poets may end, however, with a selection which is completely unaffected in spirit:

Each fountain sends up jets like the very springs of life;
The glow of the tulips makes each bed dazzling,
The morning wind uncovers the faces of the roses
And the breath of the breeze shakes drops from the roses

[6] A painter of Sassanian times.
[7] The father of Abraham, who made idols.
[8] A famous well at Mecca.
[9] "Breeze" is *saba* in Persian and hence identical with the poet's name.
[10] An *abjad*: when each letter in the final phrase is given its fixed numerical value, the date of the completion of the garden is obtained by totaling all the letters.

Into the eyes of the daffodils.
The excited shrubs sway in a light and lively dance
And scatter their golden buds over the ground.

After viewing the miniatures of the flowers and hearing how the poets sing of the gardens, it is appropriate to look at the Persians in the midst of their gardens. In following chapters Persian miniatures of gardens will be precisely related to historical periods and notable sites. At this point all that is required is a brief survey of the principal types of portrayals of gardens. It is something of a disappointment to discover that the Persian painters were never concerned with relating the features of the garden to an over all view. Bird's-eye view perspectives of gardens are completely lacking, as are scenes depicting the garden pavilion in relation to the entire garden. However, the Persian artists produced charming impressions of the atmosphere of the gardens. Such scenes were either illustrations of noted episodes in epic or romantic poetry, or portrayed contemporary monarchs and princes in their own gardens. The types were limited in number. One type depicts groups of people in a flowering landscape with a minimum of human embellishment of nature. A page from the poems of Hafiz, done in the early sixteenth century by a painter of the Tabriz school, shows a prince in an autumnal garden (Plate 9). A painting of the early sixteenth century now in the Freer Gallery of Art portrays a number of fairies dancing, singing, playing, and eating in their own garden (Plate 10). Another type of composition features a dais or fragile kiosk placed in a garden setting. Two miniatures of about the middle of the sixteenth century show variants of this type: one illustrates an episode in the *Khamseh* of the poet Nizami (Plate 11) and the other depicts a contemporary prince taking his ease (Plate 12). The last type to be mentioned is that in which all or part of an elegant pavilion or palace is shown in relation to its garden. This may be illustrated by a scene from the *Haft Awrang*, a series of seven long poems by Jami, painted in the middle of the sixteenth century, in which many elements of the garden are shown and in which a variety of outdoor activities is being pursued, some by workers and some by idlers (Plate 13). Today the Persian miniaturists paint with all the skill of the early periods and continue to turn out pleasant pictures of the gardens of long ago, but not of the gardens now in existence.

CHAPTER TWO

# Timurid Gardens: From Tamerlane to Babur

Timur—known to the Western world as Tamerlane—was one of the supreme conquerors and military leaders of the world. Around and near Samarqand, his capital city, he took delight in a gay garland of royal gardens. These pleasure places included the Northern Garden; Garden of Eram; Garden of Paradise; Plane Tree Garden; Garden of Heart's Ease; New Garden; Garden of the Glory of the World; Garden of the Image of the World; Garden of the Black Throne; Long Garden; and Garden of the Black Hill. In addition, a necklace of garden settlements named after the greatest cities of the Muslim world—Cairo, Damascus, Baghdad, Sultaniya, and Shiraz—encircled Samarqand.

Contemporary records tell that a few of these gardens remained in all their splendor for more than a century; now all have vanished. However, material descriptive of the art of the garden at this period is far more precise and detailed than for any previous period. Just as the miniatures of the Timurid period show carpets drawn and colored with exquisite detail which are older than any surviving carpets, so the gardens with their noble names endure through these same miniatures and in the words of contemporary witnesses. And the source material permits the restoration of those gardens, tents, and pavilions constructed according to imperial command.

In these accounts it is possible to observe garden design in transition, to visualize how meadows and "forests of fruit" were transformed into formal plans, and how these gardens were embellished with garden structures.

Timur, born about 1335 and descended from military leaders of a Turkish tribe, became lord of Samarqand about 1369. In 1381 he burst out of his principality to move west, into Iran. From this time on, his life was an endless tale of battles and campaigns to the Mediterranean on the west, into India on the east, and far up into Russia. Among the towns he captured were such renowned cities as Tabriz, Baghdad, Damascus, Aleppo, Ankara, and Delhi. Warfare kept him so busy that he was scarcely ever in Samarqand for any extended period. He did return to his capital in 1382, 1383, 1384, 1388, 1396, 1399, and 1404, making his longest stay between 1396 and 1398. In 1405, at the age of seventy, he died.

The historians of these remote centuries devoted most of their attention to the deeds, triumphs, and disasters of the rulers, and gave a minimum of attention to the manner in which the ruler and his subjects lived. It is, then, fortunate that three writers of the period were interested in the social background. Two of them lived at Samarqand and were well acquainted with the life of the court: one was the panegyrist Sharaf ad-din Ali Yazdi, who wrote in 1424, and another was Ibn Arabshah, who composed his work in 1436. The third author, Clavijo, was a Spaniard who went to Samarqand in 1404 as ambassador of the King of Castile and Leon: he describes royal audiences and festivities in the gardens in fascinating detail. A century later Babur (1483–1530), himself a conqueror and an emperor, and the founder of the Mogul dynasty, wrote about the verdant area around Samarqand.

Samarqand was situated in an extensive plain. Adjacent to the town was a wide zone of forests, orchards, and vineyards, and beyond was grazing land. Within this belt of green dwelt more people than were confined within the strong walls. In this belt were to be found the royal gardens and palaces and those of the powerful and wealthy feudal lords; few of the latter gardens have been recorded.

As Clavijo approached Samarqand, he noticed that "so numerous are the gardens and vineyards surrounding Samarqand that a traveler who approaches the city sees only a mountainous height of trees, and the houses embowered among them remain invisible." Throughout this green belt there was a considerable variety of landscape and vegetation. For example, about a mile outside of the town lay a meadow, possibly called the Khaneh Gul, or House of Flowers, which Ibn Arabshah described as "a carpet of emerald, on which are sprinkled diverse gems of hyacinth" and which, on holidays, was host to a tent colony. In another place, in the direction of Kesh, Timur's birthplace, lay the Green City, or Shahr-i-Sabz, in which the Palace of the Black Throne was situated.

One of the builders of the palace lost his horse in this garden and it grazed there for six months before it was found.

By combining the topographical features of a map of Samarqand drawn in 1865 with Babur's general description of the location of the meadows and gardens around the town, it has been possible to reconstruct the relative locations of these gardens in the fifteenth century (Plate 14). All the contemporary accounts speak of well-watered Samarqand. The Zarafshan (or Kuhak) River lay to the north, and numerous channels and streams traversed all the area about the town; the nineteenth-century map recorded the same watercourses.

In his personal account of the gardens Babur mentioned that they were the work of Timur and of Ulugh Beg, and he adds an account of one, the Chahar Bagh, or Four Gardens, that was created in the time of Ahmad Mirza, the late fifteenth century. It was built on a small hill and included a variety of different plots laid out in terraces and on a regular plan. Elms, cypresses, and white poplars were planted in the numerous compartments. Of the other gardens shown in the accompanying reconstruction, only the Bagh-i-Dulday, or Perfect Garden, does not appear in any of the accounts of Timur's gardens and it must be assumed that the primary concern of Ulugh Beg was in ornamenting the gardens of his grandfather with pavilions and palaces. The five meadows named by Babur supply information not included in the descriptions of Clavijo. According to Babur, these meadows were reserved as retiring places for the sultans and their families, continuing a practice of Timur's time.

The names of the streams, gardens, and meadows have survived in garbled forms or in transliterated approximations, and it is not always easy to track down the original name. For example, the meadow called Kanegil in one translation and Kanikul in another has been identified as the Kan-i-gil, or Clay Pit, but it seems just as likely to assume that its more attractive name was Khaneh Gul, or House of Flowers.

The formal relationship between the town itself and the royal garden, as Babur describes it, survived and was conspicuous at Isfahan in the seventeenth century. From the Firuz, or Victory Gate, a stately avenue, planted on each side with pine trees, led directly to the Bagh-i-Dilgusha, or Garden of Heart's Ease, across the Black Water stream.

Now that the general setting has been established, the experiences of Clavijo, the Spanish envoy, demand attention. He arrived at Samarqand in the company of the Sultan of Egypt whose presents for Timur included ostriches and a giraffe. How footsore the giraffe must have been after covering the three thousand miles from Cairo! His presence was recorded by a miniaturist at the court.

Of his initial audience with Timur, the ruler, then seventy and nearly blind, Clavijo had this to say: "We were come to a great orchard, with a palace therein . . . some distance without the city. Attendants took charge of us, holding each ambassador under his arm-pit, and led us forward, entering the orchard by a wide and very high gateway, most beautifully ornamented with tile-work in gold and blue. . . . We came to where a certain great lord of the court, a very old man, was seated on a raised dais . . . and we all made him our obeisance. Then passing on we came before another dais where we found seated several young princes, the grandsons of his Highness, to whom we likewise paid our respects.

"Then coming to the presence beyond, we found Timur and he was seated under what might be called a portal, which same was before the entrance of a most beautiful palace. He was sitting not on the ground, but upon a raised dais before which there was a fountain that threw up a column of water into the air backwards, and in the basin of the fountain there were floating red apples. His Highness had taken his place on what appeared to be a small mattress stuffed thick and covered with embroidered silk cloth, and he was leaning on his elbow against some round cushions that were heaped up behind him. He was dressed in a cloak of plain silk without any embroidery, and he wore on his head a tall white hat on the crown of which was displayed a ruby."

Clavijo indicates that this audience was in the Dilgusha garden, while another source ascribes it to the Chenar garden. However, the Spaniard mentions other occasions on which he saw Timur in state and tells of a reception in a garden held by a daughter-in-law of Timur. The noble lady and her attendants sat on a low dais, under an awning and before a great tent. Her company surrounded her, all drinking wine from gold cups, while singers and musicians entertained.

These glimpses of royalty are precisely reflected in miniatures painted at this time which depict these same royal receptions. Several representations of Timur are to be found in copies of the *Zafar nama*, or Book of Victory, by Sharaf ad-din Ali Yazdi. While none of the pictures shows exactly the scenes described by Clavijo, many of the details are the same and the atmosphere is vividly recreated. The famous double-page miniature in the Walters Art Gallery at Baltimore (Plates 16 & 17) shows Timur, with several courtiers and servants in attendance, seated on such a dais, the area adorned by fine carpets and rich awnings.

From the several references to the same gardens in the various sources it is possible to recreate features and aspects of their design. In 1396 Timur ordered that a garden should be constructed in the meadow known as the House of Flowers, a garden which he was to name Dilgusha, or Heart's Ease. Astrologers selected a propitious day and hour for the work to begin, and artists prepared the plans which would regulate the alleys and provide for the layout of flower beds. The area was plotted with perfect symmetry and provided with alleys, square beds, and "little wildernesses" of various shapes. Sycamore trees were planted along the edges of the alleys, and the remaining areas were adorned with fruit trees and flowering trees and bushes. No contemporary miniature showing the layout of such a garden is in existence, but about a century later such paintings were made and some have survived. After the garden proper was established, architects took over and provided high portals in the middle of each walled side, each side stretching some 2,500 feet. Within each corner area a small tile-covered pavilion was erected, and at the center of the ensemble a palace three stories in height was built.

It was in that same year that Timur ordered work undertaken at his Bagh-i-Shimal, or Northern Garden, so named because it lay to the north of Samarqand. Architects who had been brought from their home towns in Fars, Azerbaijan, and Baghdad entered a competition, and the ruler chose from their models and plans. This garden had the four corner pavilions and a central palace with walls of marbles and glazed tile; the walls were decorated with fresco paintings by masters from Baghdad and Iran.

As late in his life as 1404, Timur directed architects who had been brought from Damascus to begin a magnificent palace in the garden which lay to the south of the Bagh-i-Shimal, a garden which was 2,500 feet on each side. Within the palace itself were basins of water and fountains of various forms.

The garden whose plan is described in most useful detail, is that in which Clavijo was lodged in the days he was awaiting his first audience with Timur. It was near the settlement of Misr (Cairo), in that chain of gardens named after famous cities, but the name of the garden itself did not come through clearly into Spanish. It may have been the Gul Bagh, or Rose Garden. A restored plan of this garden is shown in Plate 15, and the restoration may be checked by the reader against the description of Clavijo:

"We found it to be enclosed by a high wall which in its circuit may measure a full league around, and within it is full of fruit trees of all kinds save only limes and citron—trees which we noted to be lacking. Further, there are here six great tanks, for throughout the orchard is conducted a great system of water, passing from end to end: while leading from one tank to the next they have planted five avenues of trees, very lofty and shady, which appear as streets, for they are paved to be like platforms. These quarter the orchard in every direction, and off the five main avenues other smaller roads are led to variegate the plan. . . . In the exact center there is a hill, built up artificially of clay brought hither by hand: it is very high and its summit is a small level space that is enclosed by a palisade of wooden stakes. Within this enclosure are built several very beautiful palaces, each with its complement of chambers magnificently ornamented in gold and blue, the walls being panelled with tiles of these and other colors. This mound on which the palaces have been built is encircled below by deep ditches that are filled with water, for a runlet from the main stream brings this water which flows into these ditches with a continuous and copious supply. To pass up unto this hillock to the level of the palaces they have made two bridges, one on the other part, the other opposite. . . . There are to be seen many deer which Timur has caused to be caught and brought hither, and there are pheasants here in great abundance."

Clavijo considered the Bagh-i-Naw, the New Garden, to be the finest he ever saw. The huge square area was surrounded by a very high wall with a lofty round tower at each corner. In the center was a great palace built on a cross plan with a very large pool of water in front of it. The Spaniard described another garden that he failed to name, one planted with fruit and shade trees and intersected by avenues and paths, bordered with palings, along which the guests passed. Throughout the garden were many tents with colored awnings for shade, and in the center was another of the palaces built on a cross plan.

In the above paragraphs material relating to six of Timur's gardens at Samarqand has been assembled. It is also known that the Bagh-i-Maidan, or Garden of the Square, enclosed a palace called Chehel

Sutun, or Forty Columns, which featured columns of stone and open balconies on the level of the second floor. At the time of Timur's death the Bagh-i-Chenar, the Plane Tree Garden, was under construction and a century later Babur enjoyed its pleasures.

The presence in the numerous gardens of Samarqand of similar features, indicates that they established the prototypes which were followed in gardens constructed in all later centuries. Most characteristic were the enclosure within high walls, the division of the enclosed area into quarters, the use of a main water axis, the location of a palace or pavilion at the center of the area, the choice of a natural slope or the creation of an artificial hill in order to insure proper flow of water, and a mixture of the utilitarian vineyard and orchard with the pleasure garden. Certain features seem peculiar to these gardens of Samarqand: they include the very large area taken up by the garden, the use of a square area, and the splendid portals decorated with blue and gold. The use of upper-story balconies overlooking the flowering areas seems also to have been common at Samarqand.

Clavijo and his contemporary Ali Yazdi tell of the feasts held in the gardens, but there were also frequent festivals and receptions held within much larger enclosed spaces: these enclosures were meadows and orchards which were neither walled about nor provided with permanent structures. One of the greatest of these enclosures was prepared by Timur's soldiers for the marriage of his grandson Ulugh Beg. In three or four days some 20,000 tents were pitched in irregular streets about the royal camp. Butchers and cooks passed about with their wares, bakers kept their ovens alight, and at certain places wooden cabins were built to serve as bathhouses, each with its caldrons for hot water. As the wedding day approached, the people put on festive attire, taxes and debts were remitted, and all streamed to the encampment where the "air was more fragrant than musk and the water sweeter than sugar, as though it were a part of the gardens of Paradise."

At the center of the area, the royal enclosure was surrounded by a wall of cloth some three hundred yards on a side. This wall was "as high as a man on horseback" and of many-colored silks stayed by poles within and crowned with colorful battlements in cloth. A single gateway bore above it a lofty square tower of canvas, and within was the monarch's pavilion. Clavijo's description makes it possible to visualize the appearance of this pavilion.

"It was four square in shape. In height it was the measure of three long lances such as are used by a horse-soldier, and the side was a hundred paces from angle to angle. The ceiling of the pavilion was made circular to form a dome and the poles supporting it were twelve in number each as thick round as the chest of a man breast high. These poles were painted in colors blue and gold and otherwise, and of these twelve great tent poles four were placed at the corners with two others in between on the side. Each pole was made up of three lengths which were firmly joined together to form the whole. When they had to be set up the work people made use of a windlass. . . . From each of these poles at its summit on the dome-shaped ceiling there hangs one end of a great curtain in silk cloth that is thereto attached, and these likewise form a square, and above they join to the tent wall of the pavilion. The outer tent walls of the four porticos . . . are supported by twenty-four small wooden masts . . . and the structure is stayed by upwards of five hundred ropes, and these are colored red.

"The exterior walls of the pavilion are made of a silk cloth woven in bands of white and black and yellow. Outside at each corner is set a very tall staff capped with an apple of burnished copper above which is a crescent. The summit of the pavilion further is square shaped with four tall staves at the corners, each with its apple and crescent. The staves are set at a great height and they form the framework of what is like a turret made of silk cloth set with what simulates battlements. There is a gangway from below to come up into the turret . . . to repair the faults. It is a wonder to behold and magnificent beyond description."

Clavijo was fascinated by the rich and colorful tents and described at least a dozen of them, while present interest must be more limited and related to those miniatures which show such tents within the fine gardens. Such miniatures depict the enclosure walls, the entrances to the stately pavilions, and circular tents, with those tents calling to mind the round *yurts* of the Turki and Mongol nomads. None of the miniatures depict tents with great domes, but this is quite natural since the focal point of interest in all such paintings is the human figure, and such great tents would not be in scale. Clavijo appears to describe the yurt type when he speaks of tents that are not stayed, but which have a network of poles within the canvas walls. The awnings which Clavijo saw on every hand had wooden poles at the back which gave support through taut cords, and each awning was so fashioned as to catch the breeze and provide shelter from the sun. The miniature referred to earlier (Plates 16 & 17) illustrates both the yurt type of tent and a large and elegant awning.

The miniatures of the period bring us just a suggestion of the magnitude and splendor of the royal festivals held in the gardens at Samarqand. That which centered about the vast tent described above lasted for four days, and an enormous amount of meat was washed down with quantities of wine and mare's milk. On the final day each guild of artisans put on a special show. A weaver showed a horseman and all his equipment, with all the details of armor, bow, sword, nails, and eyelids worked in fine linen cloth; the cotton weavers constructed a lofty, white tower out of this same fabric. Players, dancers, and singers wandered throughout the enclosure with their performances and jests. At several points of vantage great gallows had been erected and the populace was entertained by numerous hangings, including those of the mayor of Samarqand and of certain butchers caught charging too much for their wares. Throughout the camp trumpets sounded, decorated elephants paraded along, and fine horses with splendid trappings were seen on every hand. Red wine stained the green of the meadow, and finally Timur himself rose to dance, until from age, lameness, and drink he tottered and nearly fell. The feast ended, but its memory lingered on in the words of one who had taken part in the revels: "This life is naught but drunkenness; its pleasures recede, and intoxication comes."

Many details about the household and the administrative organization of Timur's court have been preserved. Among the important members of his entourage was a Shahab ad-din Ahmad Zardakashi, listed as a planter of trees. His name must have come to us in a somewhat garbled form: if it was Zardak kashi, he would be a planter of carrots! Of other individuals engaged in laying out the gardens we know nothing, although there are a number of references to the architects who planned palaces and pavilions. Since these architects came from greater Iran, it is natural to suppose that the Samarqand gardens reflected prevailing Persian tradition. However, features from China and India were in use. Walls were covered with Chinese porcelain, and the use of papier-mâché and interlaced patterns of wood strips reflected Chinese models, while India contributed the tradition of wainscoting in the finest marbles.

Timur was succeeded after 1405 by one of his sons, Shah Rukh, who was much admired as a peaceful, generous, and kind ruler, although these admirable qualities were not of the type most useful in holding onto the far-flung conquests of his father. Shah Rukh's wife Gawhar Shad shone as a patron of architecture, and mosques and religious schools erected at her command and under her direct supervision still stand at Mashhad and Khargird in Iran and at Herat in Afghanistan. By this time the focal point of politics and culture was moving from Samarqand down to Herat.

Shah Rukh inherited such a number of gardens that he may have felt no compulsion to add to their number. Contemporary accounts have little to say on this subject, mentioning only his interest in embellishing the surroundings of the shrine at Gazur Gah on the outskirts of Herat. The stately, polychrome-clad structures still rise proudly from a gentle hillside dotted with pools and ancient evergreens; calm and peace pervade the site.

Shah Rukh was blessed with brilliant sons. His eldest, Ulugh Beg (reigned 1447–49), was a noted astronomer who had an observatory constructed at Samarqand and who was also responsible for the creation of the Bagh-i-Nawruzi, or New Year's Garden. His brother, Baysunghur, who died quite young from trying to crowd too much into his life, was a lavish patron of the arts, a great bibliophile, and a talented painter. The cultivation of the arts proceeded apace at Herat, while the force and might of Timur's descendants waned, and provincial powers challenged central authority. A grandson of Shah Rukh, Babur Mirza, was responsible for a garden which sheltered a palace called the Tareb Khaneh, or House of Pleasure. This structure displayed a cross plan with rooms of imposing size at each side of a long central hall, and on the second floor was a series of balconies, each called a *shahneshin*, or imperial seat.

The attention devoted to the arts and sciences under the Timurids reached a crescendo during the reign of Abu'l Ghazi Husayn Bayqara who ruled at Herat from about 1468 until 1506. Babur spoke of this reign as a golden age, and has given sketchy accounts of his frequent visits to Herat and of the appearance of the embellishments of the Bayqara period.

Husayn Bayqara was crowned in the Bagh-i-Zaghan, or Garden of the Raven, at Herat, and soon thereafter ordered the construction of a chahar bagh, or quartered garden, in the open country near the shrine of Gazur Gah. Called the Bagh-i-Jahan Ara, or Garden of the World Adorned, it covered over a hundred acres and its features included a palace, pools, and masses of red tulips and roses.

In searching for precise material on the gardens of the region, it is necessary to consult the autobiography of Babur, for he not only stayed in several of these gardens, but appears to have been a conscientious tourist who visited all the palaces and gardens in and near the town. His list includes the Bagh-i-Zaghan, Bagh-i-Jahan Ara, Bagh-i-Khiaban, Bagh-i-Safid, Bagh-i-Naw, Bagh-i-Nazar Gah, Tareb Khaneh, and the garden of Ali Shir Beg. In addition to his mention of the garden of Gazur Gah, Babur speaks of a *khiaban*, or avenue, apparently the same kind of a tree-lined promenade that linked Samarqand and the gardens to the east of that town. Today the approach to the shrine of Gazur Gah has a charm and atmosphere created by its towering pines and several pools which seem reminiscent of earlier centuries, although no specific trace remains of any of the above-named gardens.

It is important to note that it was at this court of Herat and in the setting of these palaces and gardens that such renowned miniature painters as Bihzad and Shah Muzaffar did their finest work. Indeed, miniature painting had its finest flowering in this Herat school of artists and it is in the delicate, ephemeral productions of these painters that are mirrored the gardens, the palaces, and the imperial receptions that were featured in this century of culture and elegant living. In addition, the miniaturists continue to depict the life out of doors, and the tents and awnings appear as more refined versions of those shown in the paintings of Timur's period. About 1480 Bihzad did a double-page miniature of a garden of Husayn Bayqara which is preserved in the Gulistan Museum at Tehran. Another fine, unfinished painting of the same garden was formerly in the Hoffer collection.

In considering the gardens of Samarqand and Herat, an important aspect of the subject is the amount of influence that these designs and features had upon other regions and later periods. Apparently the general type migrated toward both the east and the west. Babur himself transported the type to India, where considerations of climate and topography influenced its development. To the west, the garden developed distinctive forms upon the Iranian plateau and at a later period certain of these newer features were transported to India.

Finally, after quoting Babur as a source of information concerning works of his predecessors, it is time to see what he himself did to demonstrate his extreme interest in the planning and enjoyment of gardens. In 1508 he founded the Bagh-i-Vafa, or Garden of Fidelity, on an elevated site overlooking the Kabul River and to the south of the town of Kabul. Calling it a chahar bagh, Babur writes: "In the garden is a small hillock, from which a stream of water, sufficient to drive a mill, incessantly flows into the garden below. The four-fold field-plot [the chahar bagh] of the garden is situated on this entrance. On the southwest part of this garden is a reservoir of water twenty feet square, which is wholly planted round with orange trees; there are likewise pomegranates. All around the piece of water the ground is quite covered with clover. This spot is the very eye of the beauty of the garden. At the time the orange becomes yellow the prospect is delightful."

Babur also mentioned that he had planted plane trees and sugar cane, both of which thrived in this garden. Fifteen years later he was able to return briefly to the spot: "Next morning I reached the Bagh-i-Vafa; it was the season when the garden was in all its glory. Its grass plots were all covered with clover; its pomegranate trees were entirely of a beautiful yellow color. It was then the pomegranate season and the pomegranates were hanging red on the trees. The orange trees were green and cheerful, loaded with innumerable oranges; but the best oranges were not yet ripe. I was never so pleased with the Bagh-i-Vafa as on this occasion."

It is particularly fortunate that the *Babur nama*, or the Memoirs of Babur, was so admired by his grandson Akbar, one of the great Moguls. At his court the leading artists were set to illustrate copies of the memoirs, and three different representations of this famed garden are shown (Plates 18–20). Did these artists themselves see the garden at Kabul as members of the courtly train of Akbar? Do their paintings correspond to the descriptions found in the memoirs or are they fanciful versions? In the case of the miniature by Bishandas, reproduced as Plate 18, it seems clear that he had read the words of Babur and is reconstructing elements of the actual Bagh-i-Vafa. His painting shows Babur himself instructing the architect who holds a plan of the garden. The chahar-bagh element is shown in an immensely compressed form, while the pool of water is much nearer to being in true scale. Trees bearing pomegranates and oranges border the edges of the garden. Inscribed on the miniature itself, in Persian, is a passage from the memoirs which includes a part of the first of the descriptions quoted above, and which gives the location of the garden as south of the Kuh-i-Safid Nangenhar—along the Kabul River to the east of Kabul—with *Nangenhar* being a corruption of a word meaning "good streams." Outside the entrance portal couriers press for admittance but

1. Porch of the Chehel Sutun

II. The open, octagonal kiosk which shelters the tomb of the poet Hafiz

III. View of the pool and garden of the Bagh-i-Eram

the ruler seems to have given orders that he is not to be disturbed. Plates 19 and 20 show the garden in all its glory and place special stress upon the abundance of water: Babur was accustomed to find fault with the gardens which did not have running water, and in one of these miniatures he is shown inspecting the maintenance of a watercourse in the Bagh-i-Vafa.

Babur writes about the Bagh-i-Kalan, or Immense Garden, at Kabul, which he had purchased from its previous owners and which featured great, spreading plane trees offering many shaded, sheltered, and agreeable spots. His attention turned to its perennial stream which flowed through the garden in a winding course; by altering this course to a regular one he considered that the beauty of the garden was much enhanced. This garden lay in the district now called Paghman, some sixteen kilometers to the west of Kabul, where numerous streams rush down the sides of a long, narrow valley. Along other streams Babur ordered watercourses straightened, fountains built, and circular platforms established for sitting in the open air. He gives passing mention to other gardens at Kabul: the Bagh-i-Banafsha, or the Garden of Violets; the Bagh-i-Padshahi, or Imperial Garden; and the Bagh-i-Chenar, or Garden of the Plane Tree. He relates that in the Garden of Violets he and his courtiers sat by the edge of a pool drinking wine, and then stretched out to sleep away the heat of the middle of the day.

The ruler had a continuing concern for the condition of Kabul and in 1529 he sent written instructions from India about the upkeep of its forts, mosques, palaces, and gardens. He directed that the water supplies of the gardens should be improved. At the Nazar Gah, or Place of Prospect, where Babur had established a plantation of trees, steps were to be taken to "plant beautiful trees, form regular orchards, and all around the orchards sow beautiful and sweet-smelling flowers and shrubs, according to some good plan."

Babur had a most detailed interest in the flora and fauna of the countries which fell into his hands. One example of this interest must suffice for a score. In describing the country northwest of Kabul he recorded that thirty-two or thirty-three different kinds of tulips carpeted the ground: one had a scent resembling the rose so that Babur named it *lalehgulbui*, or rose-scented tulip; another was called the hundred-leaved tulip.

After Babur settled at Agra, about 1526, he displayed his earlier interest in creating gardens, but in that region he had to cope with local problems—a scarcity of perennial streams, the prevalence of flat countryside, and climatic limitations. In nearly every site where a garden was to be brought into being, the initial step consisted in digging a deep well which might be as much as two hundred feet in diameter. If possible, a rocky outcropping or natural mound was incorporated into the plan of the garden. Unfortunately, the numerous references in his memoirs to these Indian gardens are fragmentary and in many cases the gardens are not identified by name.

On one occasion Babur chose a site near Agra which was little to his liking but on which he ordered a large well sunk. Work went forward on a great octagonal pool, a hall of audience overlooking another pool, a palace of stone, private apartments for the royal family, and a bath complex. In the end he was quite pleased and wrote that he had produced "edifices and gardens which possessed considerable regularity. In every corner I planted suitable gardens; in every garden I sowed roses and narcissus regularly, and in beds corresponding to each other." This vast, formal garden is not named but it may be identical with the site called Ram Bagh in accounts of some later years. Speculation suggests that another of the several unidentified gardens of Babur in the vicinity of Agra may be identical with the Bagh-i-Zuhara, named after a daughter of Babur, a garden which still exists. The memoirs mention a Bagh-i-Hasht Bihesht, or Garden of the Eight Paradises, near Agra and at least two gardens near Dhulphur; one was the Bagh-i-Nilofer, or Lotus Garden.

In 1529 Babur recorded his satisfaction with the grapes and melons which had been grown in the gardens at Agra; the grapes were from Balkh cuttings, and the melons produced from seeds from Kabul. However, the taste of the yield seems to have left something to be desired, since Babur states that he sent 150 porters to Kabul to bring back melons, grapes, and other fruit. The basic fact was that the gardens of Herat and Samarqand could not be transferred to the Indian plains. The climate was not suitable for orchards and vineyards, which require a cold season to establish a dormant state in the plants and the trees. In the mountainous regions the fine gardens had been the outgrowth of the bustan, or orchard, and the concept of the gulistan, or flower garden, matured at a later date. Lacking the possiblilty of producing dense, productive orchards, the Indian gardens developed toward great open spaces and wide expanses of water.

Babur was, however, able to introduce and impose upon local practice certain features of the northern gardens: the four-quarter plan, the use of running water, and formality and symmetry in planning. These features were to survive for centuries, and the irrigated garden in which water was extensively used was to become common to all northern India. Can this revolution in taste be ascribed to Babur himself? He speaks of the stonecutters, carpenters, and pool and well diggers who worked on his gardens but he never mentions an architect or a planner of these sites. In view of the intense preoccupation with gardens which is reflected in his memoirs, it seems reasonable to believe that he was his own chief designer.

One takes leave of Babur with reluctance. It is beyond the province of this work to consider the gardens of his successors, those great Mogul rulers Homayun, Akbar, Shah Jahan, and Jahangir, although it was in their later gardens that his dreams found grandiose and elegant fruition. However, we are glad to know that Babur's final wish was fulfilled and that his remains were returned to his beloved Kabul. Upon his death in 1530 his body was interred at Agra and it was not until 1597 that his grandson Jahangir managed to convey the remains to the Bagh-i-Naw, situated south and west of Kabul. An open marble pavilion shelters the marble tombstone, and in the vicinity may be found a small mosque built of white marble, a spring-fed pool, and a wood-built reception hall of recent date. Today the site is called the Bagh-i-Babur, or Garden of Babur, and it is a favorite picnic ground of the people of Kabul. Many of these picnickers know the story of Babur's attachment to the town, which he described as having "a scenery of mountains and valleys, wilderness and gardens so beautiful that the realization of this beauty completely satisfies human taste."

CHAPTER THREE

# Imperial Isfahan in the Safavid Period

One day in March 1598, Shah Abbas journeyed from his capital at Qazvin to Isfahan in order to celebrate the Persian New Year's Day, the vernal equinox, at the Plan of the World, a royal palace set in extensive gardens. In a flash of inspiration, the ruler decided to establish an imperial city in the area of orchards, gardens, and fields which lay between the famous old town and the bank of the Zayandeh River.

This development was to center around a core of palaces flanked on one side by an immense open clearing and on the other by a noble avenue, an avenue leading down to the river and designed for the pageantry of royal processions. Construction work began at once; courtiers followed the royal example and erected garden pavilions, while wealthy merchants hastened to put up shops, coffee houses, and caravanserai. By 1670 the French jeweler Chardin was able to record that the new Isfahan had 162 mosques, 48 religious schools, 1,802 caravanserai (caravan hostels), 273 public baths, and at least 600,000 inhabitants. This dedicated jeweler spent eighteen months roaming about the city, accompanied by local residents, and described its streets, squares, palaces, and pavilions in fascinating detail. From his account, from those of other foreign visitors, and from Persian sources it has been possible to construct a restored plan of the imperial city (Plate 21).

Today the visitor who lands at the Isfahan airport approaches the city along the same ground traversed in 1627 by the second British embassy to Persia. A member of that mission wrote: "We had a fair Prospect of the City, filling the one half of an ample Plain, few Buildings, (besides the High Towers of the Mosques and Palace Gates) showing themselves, by reason of the high Sycamores shading the choicest of them; yet the Hills began to keep a more decent distance, and we passed part of a spacious Field before we saluted the City; into which we entered by two fair Rows of Elms, on each hand one, planted by the sides of Crystal Streams, reaching a long way through a broad street which conducted us to the River. Here at the Foot of the Bridge many waited to welcome us, with their respective Trains, Trumpeteers, with their Ensigns, and Led Horses richly Trapped, with Pages and Attendants; thus Attended we were brought over a most Magnificent Bridge with Arches over our Heads, and on both sides Rails and Galleries. Which led us to a stately large Street, continued on the other Side with equal Gallantry of Buildings and Trees, till we were carried under their Lofty-Ceiled and Stately-Erected Bazaars."

Certainly, the elegant reception committee greeted them with *Isfahan nisf jehan*, or "Isfahan is half the world," a rhyming slogan still popular with local boosters. At their destination, the court, Shah Abbas the Great —of whom the Persians say "When the great prince ceased to exist, Iran ceased to prosper"—was accustomed to receiving English, French Dutch, and Russian diplomats, and was not surprised to learn that members of the British embassy found that Isfahan surpassed London in magnitude and splendor.

The focal point of the royal plan was the Maidan-i-Shah, or Imperial Square (Plate 22), established on the site of an irregularly shaped plot of ground used only on market days. Converted into a formal rectangle, 570 yards long by 175 yards wide, it remains one of the largest public squares in the world. Each side of the square presented a uniform façade two stories high, housing the shops of goldsmiths, dealers in precious stones, druggists, and merchants of cotton, woolen, and silk cloth—some of these same businesses still function in these shops.

A number of rather narrow streets, replaced within recent years by wide, paved avenues, converged from the older town into the square, which was the scene of hectic activity on weekly market days. Under hundreds of temporary tents and awnings dealers sold jewelry, pearls, and furs; petty traders offered cloth, spices, vegetables, and fruit; and ladies of easy virtue drained away the cash sales of farmers from nearby villages. Jugglers, tight-rope walkers, acrobats, magicians—one of whom performed the Indian rope trick—and wrestlers competed for the small change of the crowds. At night the square was lighted by 50,000 tiny lamps hung from slender poles. Criminals might receive public punishment. On one occasion a baker who had sold underweight loaves in a period of scarcity was seized and brought to the square. A fire was kindled and nourished to a roaring blaze over which a great copper tray was heated to a red glow: the executioner struck off the head of the kneeling baker and, taking it by the hair, placed it on the tray. Nerves and blood vessels being immediately singed and sealed, the head remained alive long enough to weep and wail.

The climax of the spectacles in the square consisted in those events in which the Shah and his courtiers took part. Fine horsemanship was featured and some of the feats may recall those of our rodeos. A courtier would gallop down the field throwing to the ground twenty small plates and then, galloping back at full speed, snatch up all the plates. Archers shot at a ripe melon or a bag of gold coins suspended by a cord, with the archer cutting the cord receiving the money. Or, in the words of a spectator: "[For] the Exercise of the Bow on Horse-Back . . . the Gentleman takes his Career towards the Pole, bearing a Bowl or Cup, with his Bow and Arrow in his Hand, and when he has gone by it, he bends himself backwards either to the Right or Left, for they must know how to do it both Ways, and lets fly his Arrow."

However, the most regal of these events was polo, native to this part of the world. A seventeenth-century account paints this picture: "So the King went down, and when he had taken his horse, and the trumpets sounded: there were twelve horsemen in all with the King: so they divided themselves six on the one side, and six on the other, having in their hands long rods of wood, about the bigness of a man's finger, and on the ends of the rods a piece of wood nailed on like unto a hammer. After they were divided and turned face to face, there came one into the middle, and did throw a wooden ball between both the companies, and having goals made at either end of the plain, they began their sport, striking their ball with the rods from one side to the other and ever when the King had gotten the ball before him the drums and trumpets would play one alarum." Two pair of marble goal posts still stand, one pair at each end of the square.

In 1612 the construction of the Masjid-i-Shah, the Imperial Mosque, was begun at the southern end of the maidan. Spurred on by the stern commands of Shah Abbas, referred to in one of the inscriptions in this mosque as "the most noble in lineage of the sovereigns of the earth, the most honored because of his personal valor, the greatest by rank and position, the most cognizant in argument and proof, he who best unites justice and mercy . . . ," the workers raised its walls rapidly. Initial attention was given to the monumental portal facing onto the maidan, and all the details of the complex structure were not completed until twenty years of effort had been expended.

This mosque is one of the supreme monuments of Persian architecture. Inside and out the surfaces of walls, pillars, vaults, and domes are clad in multicolored tile in which tones of light blue and dark blue predominate. The lofty entrance portal, flanked and crowned by a pair of minarets, leads into a spacious, open central court, surrounded by arcades in two stories. Across the court from the entrance portal another great arch leads into the sanctuary of the mosque, a vast, square chamber crowned by a soaring dome (Plate 23). Because of the royal insistence upon haste, the mosque rose on inadequate foundations and in later centuries earthquakes split great cracks in the walls. Over many years the Iranian authorities have labored to consolidate the fabric with steel and concrete, while patient craftsmen have replaced millions of pieces of missing tilework. Today the monument is as impressive and beautiful as when first completed.

Corresponding to the portal of the Imperial Mosque, a monumental entrance to the bazaar was erected at the northern end of the maidan in 1617. Its façade was decorated with a representation in colored tiles of the constellation Sagittarius, and on the rear wall was a tremendous painting of a victorious battle of Shah Abbas against the Uzbeks—areas of this painting are still visible. The portal itself was named the Naqqara Khaneh, or Drum Tower, because it was crowned by a gallery where "their music, composed of kettled drums, oboes and other instruments, is heard every evening at sundown and when the Shah enters or leaves the Maidan." This practice of drumming the sun to rest, far older than the Safavid period, is still practiced in Iran.

Behind this portal a royal bazaar, crowned with vaults and domes, was erected to connect with the long lanes of the covered bazaars of the older town. In this area "are sold the richest stuffs and goods that are found in all the realm and it is so full of shops for all kinds of merchandise that there is nothing in the world so rare that it may not be found there." This bazaar remains a magnet for visitors to Isfahan: separate sections are given over to coppersmiths, to the jewelers, and to the rug dealers, while numerous, crowded shops display the characteristic hand-blocked cotton prints of Isfahan, colorful local pottery, hand-loomed silks and wools, and mounds of nuts, raisins, and sticky sweets.

Clustered along the bazaar were the caravanserai—both inns and wholesale warehouses—and the quarters of the surgeons and barbers, as well as taverns and cabarets, including tea and coffee houses. In 1627 an Englishman wrote: "Their coffee houses are great, lofty halls and are among the most handsome buildings in the city because they are the meeting and amusement places of so much of the population. In the center of the hall is a huge basin of water and around the edge are wide benches on which they sit cross-legged in the Oriental mode. People talk over news and politics and play at a game like checkers, while poets make their rounds, reciting their verses, and holy men deliver impromptu sermons. The coffee is very properly served, very quickly and with great respect."

Coffee had been unknown in Europe until travelers returning to Europe from Turkey and Iran reported on this strange drink. One of them wrote: "Chawa is made from a fruit . . . which has the taste of Indian corn and is the size of a horse bean. They roast it on a stove, then make it into a powder and then boil it with plain water; and thus make this drink which smells burnt and is not pleasant to drink. But the Persians believe that this water is capable of extinguishing the natural tendencies toward amorous passion and so when they do not wish to be overburdened with children they drink it frequently." This opinion was current even in the harem. One day a lady of the royal household gazed from her apartment overlooking the stables onto a bustle of activity. Asking a servant what was the matter, she was told that the grooms were trying to exercise a stallion belonging to the Shah but that the animal was so fierce and strong as to be unmanageable. The lady pondered a moment and then said: "That's easy enough! All the grooms have to do is give the horse coffee to drink, for since the Shah has taken to drinking it he is of no use at all to us women."

In this same period the Persians were addicted to tobacco, although their doctors warned them that smoking dried out the body and thinned and weakened the frame. Shah Abbas himself was alarmed at the headway the habit had made and tried in vain to banish its use. One day, when all his highest nobles were present, he ordered water pipes passed around which contained dried and shredded horse dung instead of tobacco. As they puffed away the Shah asked: "How do you like this tobacco? It's a present from my governor at Hamadan, who claims that it is the best in the world." Each replied in turn: "Marvelous." Finally, the Shah turned to his favorite general and said: "Tell me the truth, what do you think of this tobacco?" "Sir, on your sacred head, I swear it smells like a thousand flowers," came the reply. The Shah roared: "Cursed be that tobacco drug, which you can't tell from horse manure."

At the center of the long, east side of the maidan is situated the well-preserved mosque of Shaykh Lutfullah. This was one of the first structures built, for it was partially completed by 1603, although the decorative details were not finished until 1619. From its faïence-coated façade a corridor leads to the single room of the mosque, a square chamber transformed at a higher level into an octagon upon which rests the dome. Every square inch of the interior surfaces of the walls and vaults is covered with faïence mosaic—myriads of tiny segments of multicolored glazed tiles set together to form elaborate patterns. Pierced grilles in the sixteen windows of the dome produce changing patterns of light pouring upon the sparkling walls. The effect is so elegant, so rich, and so dazzling that for minutes at a time the spectator loses all touch with time and reality. Even the Arabic inscriptions are a delight of pure decoration, and properly so, for they were designed by one of the most renowned miniature painters of the period. On the exterior of the dome are swirling patterns of floral abaresques in blue, green, and white upon a background of tawny-colored tiles (Plate 24).

At the center of the western side of the maidan, opposite the Shaykh Lutfullah mosque, stands the Ali Qapu, or Lofty Gateway—the noble entrance to the palace grounds and gardens (Plate 25). In the earlier stages of building the imperial city, it served the double function of an entrance to the gardens and a center for the administrative offices of the court. The type of structure was most unusual for Iran—a veritable skyscraper rising six full stories and with as many as ten rooms on some of its floors. The upper stories were used by Shah Abbas himself and were richly decorated. On the ceilings were floral patterns in red, blue, and gold; on the walls scenes of courtly pleasures in garden and in country, portraying "men and women standing and drinking, armed with bottles of wine and glasses in their hands and also in many other postures which were intended to signify that here Bacchus and Venus were perfectly matched." The top floor included porcelain rooms. Walls and vaults were lined with thin wood and plaster panels, cut into seemingly fantastic shapes. Actually each such opening was a niche designed to hold a special shape of vase, bowl, or jar to display the fine Chinese porcelain which Shah Abbas collected in quantity.

From his rooms on the fourth floor of the palace the Shah could walk out onto a lofty open porch which overlooked the entire maidan. Eighteen huge tree trunks sheathed with strips of wood still uphold the roof, with its ceiling decorated in geometric patterns of blue and gold. Set into the floor of the porch is a marble basin, lined with lead sheets: three fountains once played in this basin, fed by water forced up by an ox-powered machine. Today the visitor ascends to the very roof of the Lofty Gateway to enjoy an unobstructed view over the sparkling city, hemmed in by emerald fields and by rose and blue mountains.

Royal audiences were held at the Ali Qapu and the reception of foreign ambassadors took place in a blaze of splendor. Grouped before the gateway was a varied display: ". . . twelve of the Shah's finest horses, six on each side, harnessed with bridles inlaid with emeralds, rubies, and gold enamel. The saddles were strewn with precious stones and each wore a cloth of gold brocade covered with pearls. The horses were tethered by ropes of silk and gold to two stakes. Thirty paces from the horses were savage beasts reared to fight with young bulls, two lions, a tiger, and a leopard, each lying on a fine large rug. Before them were two great basins of gold from which they fed. Near at hand were two gazelles and to the left two elephants crowned with canopies of gold brocades. There was also a rhinoceros." The ambassadors ascended to the porch, with their attendants bearing presents for the Shah. The Russian ambassador offered a huge glass chandelier, mirrors with painted frames, fifty fine furs, and twenty bottles of vodka. The ambassador from the king of Bosra presented an ostrich, a young lion, and three fine Arabian horses. As the gifts were presented, drums and trumpets sounded and combats of wild beasts began in the maidan.

Up on the porch, feasting was in order: "First a refreshment consisting of fresh and dried fruits, jams, and liquids was brought around. It was served in great lacquered trays, each of which held a nest of twenty to thirty little porcelain dishes. At the end of the porch was a huge buffet, one side holding fifty great bottles of wine and on the other scores of cups. The Shah and his nobles drank draughts of wine and the Russian ambassador vodka." Unfortunately, on this occasion the Russian was not able to hold his liquor. Forced by urgent necessity, he snatched off his tall fur headgear and held it in his lap as a basin. That was bad enough but worse came when his attendants helped him to take his leave, for he thrust his hat back on his head to the mingled delight and disgust of the company.

From the central archway of the Ali Qapu a corridor led west to the palace grounds proper, passing the barracks and the mosque of the royal guard. On the left was the Talar-i-Tavileh, or Hall of the Stable, so named from its location in the vicinity of the royal stables housing one thousand horses. This structure vanished in ruins more than a century ago and it is fortunate that an engraving made before 1700, and illustrating a splendid night reception, has survived (Plate 26).

Building and rebuilding was the rule within this area, and in this same section a structure called the Sar Pushideh was erected, reportedly in the opening years of the nineteenth century at the orders of Said ad-daula Mirza, a son of the Qajar ruler Fath Ali Shah (reigned 1797–1834). It did not last out the century. The name may be translated as the Covered Head, or Covered Porch, and appears to stem from the fact that the part of the building which was normally left open to the elements was here closed in and roofed over. The illustration shows how a typical garden pavilion was turned inside out; it was probably intended for occupancy during the colder months (Plate 27). One of the column bases shown in this old drawing now stands in the garden of the nearby Chehel Sutun, or Palace of Forty Columns. The drawing shows features of design and decoration which suggest that the pavilion may actually have been erected in the eighteenth century and merely repaired at a later date.

Further along and on the right a considerable area was given over to royal storehouses and workshops. Separate buildings served as warehouses for court robes of honor, for torches, for coffee, for pipes, for wine, and for a variety of foodstuffs. The library and a workshop for illuminating manuscripts and binding books were housed in one building; other workshops held the weavers of precious brocades and fine rugs, as well as the European watchmakers and jewelers employed at the royal court.

Few were privileged to penetrate beyond this area—only the principal nobles, invited guests, and servants and guards. For these favored few and for the royal household, the inner enclosure offered a delightful expanse of woods and gardens, pools and fountains, and countless blue-tiled channels of running water. In one section of the enclosure four charming pavilions were scattered among the trees: these structures—the Guest House, the Building of Paradise, the Hall of Mirrors, and the Building of the Sea—are no longer to be seen. According to one account, such structures were "made expressly for the purposes of love. The furnishings of each part are the most magnificent in the world and the most voluptuous, with retreats which are nothing but an entire bed." To the south of these pavilions lay the extensive complex of the harem or serail—the residence of the Shah's wives and concubines. The most elegant of the structures in this complex housed the Shah and his thirty favorite wives, each such wife having two rooms for her private use. Adjacent were dormitory-like buildings with 150 separate apartments; eight or nine women were in each second-floor apartment, with eunuchs and female servants living on the ground floor. In the more distant reaches of this area separate structures housed the numerous children of the Shah, as well as the elderly women of the royal family. Not one of these buildings has survived the centuries, and the contemporary accounts shed little light upon the secret life of the harem. One seventeenth-century traveler noted that "women are more closely guarded in Persia than anywhere else on earth. This is because the climate of the country is so hot and dry that it causes an extremely violent passion for women." Male jealousy could be a terrible thing: there is the story about a resident of Isfahan who flayed and skinned his wife alive just because she had looked at another man. He then stuffed the skin with straw and hung it outside of his door as an object lesson for potentially flighty females.

Officials of the Shah were continually seeking out the most beautiful girls of the country and despatching them to the harem at Isfahan. Indeed, local beauty contests were held with the winners receiving one-way tickets to the court; naturally, only virgins were considered for this honor. Once settled in the harem a woman usually passed her life there, although not infrequently the Shah would present one of the attractive inmates to a favorite noble who responded by marrying the girl. Scores of black or white eunuchs were constantly on guard to repel intruders, and to attempt to preserve internal order among the plotting, bickering community.

Located at the opposite end of the grounds from the harem is the Chehel Sutun, a building which has survived in excellent condition and which is a highlight of a visit to Isfahan. The plan of this palace pavilion (Plate 32) reflects Persian ingenuity in that free outdoor space and enclosed living space were brought into such an intimate relationship that it is hard to say where one stops and the other begins. This concept is just beginning to play an important role in residential architecture in the United States. At the front of the pavilion a great porch with towering wooden columns overlooks a long reflecting pool, and the ceiling of the porch is made bright with painted wood mosaic, inlaid with mirrors of many shapes (Color plate 1). At the center of the porch four massive stone lions spout water into a marble basin (Plate 28) and in the recess at the back of the porch the royal throne was once conspicuous. The wall behind the throne is sheathed with mirrors, which serve to give this part of the structure an airy, unsubstantial character.

From the porch of the Chehel Sutun three doors lead into a banquet hall which extends the full width of the pavilion. Six immense oil paintings adorn the upper walls, and help the spectator to visualize the turbulent and lusty life of the seventeenth century. One depicts Shah Abbas playing host to the khan of the Uzbeks. The two kings are seated on a dias, with Shah Abbas, wearing a calico turban and flaunting a handlebar mustache, holding out his cup for more wine. In the background servants bring in golden platters, and in the foreground dancing girls flourish castanets and tambourines. Lucky European travelers of the period delighted in writing about these royal parties: "Then the dancing wenches went to work, first throwing off their loose garments or vests—the other was close to the body resembling trousers, but of several pieces of satin of sundry colors; their hair was long and dangling in curls; about their faces hung ropes of pearls, and about their wrists and legs were wreathed golden bracelets with bells, which, with the cymbals and timbrels in their hands, made the best concert. Their dancing was not after the usual manner, for each of them kept within a small circle and made, as it were, every limb dance in order after each other, even to admiration. This kind of dance being so elaborate that each limb seems to emulate—yea, to contend—which can express the most motion: their hands, eyes, and bums gesticulating severally and after each other, swimming round and now and then conforming themselves to Doric stillness."

This banquet hall now houses a museum of Persian works of fine arts, while ancient panels at the level of the dado repeat the scenes of rustic enjoyment common to the Ali Qapu.

To the west of the entire palace area was the avenue of the Chahar Bagh, or Four Gardens, which was not planned to serve as a city street —it lay outside the more densely populated area—but as a promenade. Beginning at the northwest corner of the palace grounds, it ran down a slight grade for nearly a mile to the river, across a splendid bridge, and then up rising ground to a vast royal estate called the Hazar Jarib, or

Thousand Acres. Eight rows of plane trees and poplars, among which grew a profusion of roses and jasmine, were spaced across the sixty-yard width of the promenade. Most of these trees were planted in the presence of Shah Abbas with a fertilizing deposit of gold and silver coins at their roots. Five watercourses ran down the avenue, the wider central one was lined with cut stone, and breaks in level were marked by pools and fountains, each pool of a distinctive design.

Chardin wrote that it was the most beautiful avenue he had ever seen or heard talked about, and it retained all its early charm until near the end of the nineteenth century when a son of the ruler, in his post as governor, had many of the fine old plane trees cut down. In recent years the avenue has become a main traffic artery, paved with asphalt. Had someone in high position displayed enlightened concern, the original atmosphere could have been preserved. In addition, the maidan itself has been desecrated by the intrusion of a pool and a fenced-in flower bed.

On either side the Chahar Bagh was lined with attractive structures. At the head of the avenue stood a three-story pavilion behind whose latticed windows the principal ladies of the harem assembled to witness the ceremonial arrival of ambassadors and the stately excursions of the royal household. Successive gardens along the way bore descriptive names, such as the Octagonal Garden, the Garden of the Nightingale—the sweet-voiced nightingales of Isfahan still pour their melodies on the evening air—the Mulberry Garden, and the Garden of the Vineyard. Each garden contained two pavilions, a small one built over the entrance gate and a large one at the middle of the enclosure. Lattice-work walls along the avenue did not shut out the view and the passersby could see fine carpets spread on the ground and clusters of fresh flowers floating on the pools. One of these vanished garden palaces is shown as it appeared in the nineteenth century, well in ruins (Plate 29); it was the Bagh-i-Zereshk, or Garden of Barberries.

To the west of the Chahar Bagh, in the section called Asadabad, was a garden pavilion in the style of the Chehel Sutun. A sketch made of a diplomatic reception at this pavilion prior to 1700 is of particular interest, since many of the details—such as the horses and lions on display—correspond with the description given earlier of a reception at the Ali Qapu (Plate 30).

At the point where the avenue reached the Zayandeh River, quays stretched along the bank. At one corner angle of avenue and quay was the royal aviary, featuring a huge cage meshed with gilded wire, and at the other corner the royal lion house, which also sheltered bears, leopards, elephants, and a rhinoceros. How many weary miles from the Persian Gulf this same rhinoceros trudged over burning roads and up steep passes, in low gear, before he reached Isfahan!

Of all the airy palaces and pavilions built along the Chahar Bagh only the Hasht Bihesht, or Eight Paradises, remains. It was erected in the vicinity of the earlier Garden of the Nightingales during the reign of Shah Sulayman (1667–94) and was seen by Chardin in 1670. The restored plan shows the location of the pavilion within its own spacious grounds (Plate 33); today its battered shell is hemmed in by modern shops. The plan was a customary one but displayed an unusual treatment of the corner angles, as illustrated in a perspective drawing made a hundred years ago (Plate 31). Particularly charming was the view from the central hall; the superstructure is related to the domed tombs of the period but the lantern and the portals are opened wide to the exterior and the decoration is as gay as possible (Plate 34). On the second floor the fabric was pierced with galleries and chambers. Each little chamber had its special shape and decorative scheme; some had basins with fountains fed by lead pipes embedded in the walls, and some had ceilings and walls entirely lined with mirrors. Most looked out over the garden and its several pools.

Just to the south of the Hasht Bihesht is a structure now known as the Madrasa Mader-i-Shah, or the Religious School of the Mother of the Shah, which was completed in 1714. The entrance façade on the avenue presents a brilliant sheen of multicolored tile and polished marble. Within, a spacious court covers an area earlier taken up by the dwellings of the king's children. Around the rectangle of the court are two-story arcades which open into separate living rooms for the religious teachers and their students (Plate 35). A great dome rising above the sanctuary chamber delights the eye (Plate 36). The lush green of the trees, the hues of flowers against the bright colors of the tilework, the reflection of the whitewashed arcades in the shimmering, dark pool: all these striking contrasts dazzle the senses.

The bridge that carries the Chahar Bagh across the river, built for Shah Abbas by the general of his armies, now bears heavy truck traffic as well as an occasional camel caravan. The sturdy structure, almost

IV. View of the main section of the Divan Khaneh

V. Typical smaller town house and garden at Shiraz

three hundred yards long, excited the admiration of such an experienced observer as Lord Curzon who wrote: "One would hardly expect to have to travel to Persia to see what may in all probability be the stateliest bridge in the world." Sections of the bridge are clad with faïence, while deep galleries in its massive piers are frequently visited by picnickers who settle down to enjoy the cool shade, the rushing water, and the fresh breeze of the river valley. The bridge can be seen in the background of Plate 37.

On the south bank of the river and to the east of the extension of the Chahar Bagh was a royal garden which extended as far as the Khaju bridge. The area was known as Sa'adatabad, or the Abode of Felicity, and was described in two works by contemporary poets—*The Rose Garden of Prosperity* and *The Secret Language of Sweet Smelling Flowers.* The most important structures in this garden were the Haft Dasht, or Seven Perfections; the Ayina Khaneh, or Hall of Mirrors; and the Namakdan, or Salt Cellar. To the east of the Haft Dasht, near the end of the bridge, was the Ayina Khaneh, while the Namakdan was behind and to the south of the Ayina Khaneh.

The Haft Dasht comprised an impressive complex of structures arranged around an open court and surrounded by high walls, most apparent along the river bank. The women of the royal court were long housed in the Haft Dasht, but by the early nineteenth century all the buildings had become quarters to house visiting Europeans and other dignitaries.

The Ayina Khaneh was also called the Divan Khaneh, or Audience Hall. The illustration shows it as it appeared in the middle of the nineteenth century, with the Siosehpol Bridge in the background (Plate 37). The building was destroyed at the end of the century. With the *talar*, or great porch, placed in front of a series of rooms, the general arrangement was very like that of the Chehel Sutun. One of its column bases, decorated with figures of lions in relief, now stands in the garden of the Chehel Sutun.

In back of the Ayina Khaneh an avenue led to the nearby Namakdan. Since the structure was octagonal and rather tall, it was frequently referred to as the Kola Ferangi, or European Hat. Elevated in two stories, it was an airy, open building, noteworthy for the fact that jets of water surrounded it with a solid sheet of spray. About 1885 it was torn down by a local prince in search of building materials.

To the west of the area where the Chahar Bagh spanned the river, was (and still is) the village called Julfa, where Shah Abbas settled several thousand Armenians brought down from Julfa in northwestern Iran to do weaving and embroidery for the royal court. Today the visitor may marvel at the Armenian cathedral with its fine mosaics and sacred paintings, and may move directly into the past in the museum just adjacent to the cathedral.

The Chahar Bagh continued up rising ground to end at the entrance to the Hazar Jarib, a garden laid out about the middle of the seventeenth century. According to Chardin, it was a mile square and arranged in twelve terraces, each a few feet higher than the ones nearer the river. Twelve avenues ran parallel to the direction of the Chahar Bagh and there were three east-west transverse avenues. Along every fourth avenue was a stone-lined canal, with a basin of a different shape on each terrace. Water jets were everywhere, and there were at least five major pavilions. In the spring the entire garden was carpeted with flowers, especially along the canals and around the pools, and "one was surprised by so many fountains appearing on every side as far as one could see, and was charmed by the beauty of the scene, the odor of the flowers, and the flight of the birds, some in aviaries and some among the trees." Much of the construction was in mud brick and only lines of low walls remain today.

To the west of the Hazar Jarib and on higher ground well back from the river was an even more extensive royal garden known as Farahabad, or Abode of Joy, which had been built to the orders of Shah Husayn about 1700. It lay a good long horseback ride, nearly ten kilometers, from the maidan at Isfahan.

About 1930 a French architect strolled back and forth across the site, sketching lines of walls, plotting the locations of islands in vast reservoirs, and establishing the major and minor axes of the original plans. His drawings do much to bring the ephemeral site back to life—the site where hundreds, if not thousands, of servants once waited upon royalty. The extensive service area, studded with dozens of courts and separate units, was on the highest ground, and from there to the river on the north the distance was about the same as that from the Washington Monument to the Capitol building. Huge groves of trees and plots of orchards filled the area, as well as two exceptionally large daryacheh, or little seas, each with a summer house on an island.

Contemporary accounts indicate that the life of the rulers at Isfahan was a steady round of visits from the main palace area near the maidan to Sa'adatabad, to Farahabad, and, later in date, to Hazar Jarib. Each such move was thought of as an escape from the cares of state, and each such outing entailed the movement of women, courtiers, servants, equipment, and supplies. However, it would be incorrect to picture the lives of Shah Abbas and his immediate successors as featuring carefree ease. Each ruler had to move about the entire country to administer justice, to visit sacred shrines, to pacify unruly provinces, to enlist tribal support, and to lead his troops against foreign and domestic enemies. The periods of pleasure spent in the gardens of Isfahan were hard won and thoroughly deserved.

This picture of Isfahan in the seventeenth century is far from complete, but the town itself remains in all its unique and transcendent character, awaiting visitors from the other half of the world.

CHAPTER FOUR

# Gardens along the Caspian: Safavid and Later

Along the coast of the Caspian, the Persian gardens are set within an entirely different climate and topography from that prevailing on the lofty and generally arid, barren plateau. This coast is as much as one hundred feet below sea level, and stretches along a salt sea. It receives forty to sixty inches of rain a year and is covered with marshes, jungle, and intensive cultivation.

The first sight of the region is almost overwhelming, especially to the traveler who follows the shortest route from Tehran. The road winds along barren slopes and up narrow valleys to cross the height of the Elburz Range at about eight thousand feet. On a good day the view northward across the green-clad slopes and far over the Caspian is spectacular; patches of forest and sea may emerge from a blanket of clouds some thousands of feet below. Down the road descends, to continue through great woodlands, past the settlements of charcoal burners, and, finally, to emerge on the cleared coastal strip. Here rice is the major crop, with tea plantations scattered at intervals. The atmosphere is tropical and the architecture as well, for many houses stand well above the ground level on stout wooden piers and display steeply pitched thatched roofs.

The coastal strip to the west has long been known as Gilan, and that to the east as Mazanderan; most of our attention will center upon Mazanderan. As we are to consider older gardens, it is well to quote descriptions of the region made by travelers who saw these gardens in their prime. Jonas Hanway, who visited Mazanderan in 1744, wrote: "They enjoy here a long spring; their lawns and meadows are strewn with flowers, and the bushes with honey-suckles, sweet briars, and roses. The soil is exceedingly fertile, producing . . . almost every kind of fruit without culture; for besides oranges, lemons, peaches, and pomegranates, here are abundance of grapes . . . growing wild in the mountains with great luxuriancy; so that a considerable part of this province is quite a paradise. . . ." However, Hanway felt compelled to say that the air of the region was extremely moist and far from healthful, being productive of agues and of numbers of frogs, nits, and spiders. He did note that, "Women, mules, and poultry enjoy health, when all other animals pine away with sickness." Until a few years ago malaria was the scourge of the population but the repeated spraying of thousands of houses and ponds has done much to eliminate the incidence of malaria in extensive sections along the coast.

Hanway pointed out that Shah Abbas delighted in the southern coast of the Caspian and particularly in Mazanderan, and it is with his palaces and gardens that we are to be concerned. According to report, Shah Abbas built a lodge every two leagues to provide for his rest and refreshment when traveling along the coast. It is certain that he built a splendid road on which to make these travels, and this causeway stretched for nearly three hundred miles. The road was raised high above the level of the damp or flooded ground, was ditched on either side, and measured from twenty feet to as much as twenty yards in width; it was paved throughout with large water-worn boulders from the streams. Because it was never repaired, this causeway fell into ruin, although parts of it may still be seen; stretches of modern motor roads overlie some sections of it.

Six garden sites of Shah Abbas and one of Shah Reza Pahlevi, who came to the throne some three hundred years later, remain to be considered. All are in Mazanderan and, proceeding from west to east, the sites are Amol, Barfurush, Sari, Farahabad, Safiabad, Ashraf, and Abbasabad (Plate 39). The value of examining these sites is to identify the plan forms and design features in use along this coast where the climate is so different from that of the plateau. It will be apparent soon that nearly all the features are common to those of the palaces and gardens at Isfahan.

At Amol a palace built by Shah Abbas survived for at least a century and a half after his time but has now vanished. It was two stories in height, of stone, commanded a pleasant prospect, and was carefully contrived for coolness and comfort. As for the garden in which it stood, we know only that it was remarkable for the size and height of its cypresses.

Near the settlement of Barfurush, now known as Babol, Shah Abbas ordered a garden laid out which did not conform to the customary types. South of the town a stream was led into an area enclosed by an embankment, so that a lake of nearly two miles in circumference was created. In the lake was an artificial island of some two acres in size, and brick pillars supported a long wooden bridge from the mainland. Other pillars within the lake indicate that at least one pavilion stood by itself in the midst of the lake so that it could be reached only by boats. This garden was known as the Bagh-i-Shah (Royal Garden), or as the Bahr ol-Eram (Sea of Paradise). Today the Safavid buildings have all vanished and there are remains of only two nineteenth-century brick structures, a residence with *anderun* (women's quarters) of Mahmud Quli Mirza. Even in the absence of the early buildings, it is possible to visualize the garden of Shah Abbas as a variation of the daryacheh, or "little sea" gardens, common to the plateau proper. In those gardens hillside springs flow down into a vast pool, but on the Caspian slope water is so much more abundant that the pool, as at Amol, becomes a real lake.

In or near the important town of Sari are the remains of three gardens, two of them associated with Shah Abbas. To the north of the town is his Bagh-i-Shah, which today displays, on the edge of a pool, a garden pavilion repaired at some later date with a roof of red tiles. Two gardens were situated along the maidan of the town. One, on the site of a garden of Shah Abbas, was constructed as the palace of the local governor in the reign of Aqa Muhammad Qajar (1779–97) who himself enjoyed staying there. European travelers of the nineteenth century speak of state apartments decorated with paintings showing the victories of Shah Ismail (reigned 1500–24) and Nadir Shah (1736–47), and as that century drew toward its close these accounts reflect the neglect and gradual decay of the structure. However, Hommaire de Hell has left a view of the palace and its garden as it appeared in 1847 (Plate 38); it is interesting to note that several travelers spoke of the proximity of plane trees and date palms in this garden. The other garden, facing that of Aqa Muhammad Qajar across the maidan, was established by a local governor, Muhammad Quli Mirza.

In the fall of 1611, Shah Abbas ordered construction begun at the village of Tahan on the banks of the Yijan near where that river emptied into the Caspian, and gave the site the name of Farahabad, or Abode of Joy. That it was an extensive development may be learned from a description made some two hundred years later by James Fraser. From his words it is apparent that the area was developed along lines similar to that of the maidan and of the surrounding structures of Shah Abbas at Isfahan. At Farahabad the maidan had its long axis from north to south, and was two hundred meters long and more than one hundred meters in width. On the northern side, adjacent to the Caspian, was a vast walled enclosure divided by a high wall into two sections: one for official receptions and court functions, and the other a private residential area for the ruler. Baths and service structures were near the maidan while in the private area and close to the sea was the palace called Jahan Numa, or Image of the World. To the south of the square rose an imposing mosque crowned by a dome. The maidan itself was arcaded on every side, so that a later traveler described it as a vast caravanserai. Shops gave onto these arcades, and to the east and west of the square a town, with the sole function of serving the court, came into being. At the same time, work was started on a causeway toward Sari, a distance of some thirty kilometers.

Work was rushed on the many buildings, and when the development shone in its freshness and splendor an Englishman, Thomas Herbert, visited the palace in 1628. He wrote: "[the] sumptious palace of the Empereurs at the north side of the City . . . extends to the Caspian. It has two large courts (comparable to the Fountain bleaus) either of them which expresse an elaborat Art in the skilful gardiner . . . the spreadings Elms, Chenores, and Sicamores surrounding. . . . The house is low, but each chamber, high and spacious, rich in work . . . some are square, some gallery wise, but all arched: three were especially rich and lovely; whose sides were set with Mirrors or Looking-glasses and whose tops or ceilings were gloriously embost with flaming Gold; the casements were of large square Muscovian glasse, cemented with Gold . . . the ground was overspread with crimson velvet, some stuft with Down . . . in those galleries of Mirrors the King has sundry representations of venereous gambolls, his Concubines studying by amorous postures to illure his favor, to glut his fancie; the other chambers are richly furnisht, the walls varnisht and painted in oyle, but by an uncivill pencil, the genius of some goatish Apelles; such Lavaltoes of the Persian Jupiter are there, such immodest postures of men and women, nay of Paederastyes, as makes the modest eye swell with shame. . . ." As useful as this account may be, we wish that Herbert had given closer attention to the garden and the arrangement of the buildings within it, rather than to the paintings which so shocked him. Actually Herbert was shocked everywhere he went—at Farahabad, Ashraf, and Isfahan—but he always stops just short of telling exactly what he saw.

Another of the valuable sources for this period, Chardin, visited the palace in 1666, and his jeweler's eye was attracted by things other than paintings. He admired, "a vast Treasure of Dishes and Basins of Porcelaine or China, Cornaline, Agate, Coral, Ambar, Rock Crystal. . . . [They] built the great Haouse or Tangi, being a Jasper Fountain covered with plates of gold and erected within the palace." Chardin is speaking of a *hauz* or *tang*, both Persian words for a pool, situated in a main room of the palace.

The palace seen by Herbert and Chardin was the Jahan Numa, and from later accounts we can be sure that it was similar in plan and scale to the palace of the Sahib Zaman, or Lord of the Age, at Ashraf. Its crumbling ruins were seen and sketched until well into the nineteenth century, although its decline began as early as 1668 when a horde of Don Cossacks landed and sacked the site.

Shah Abbas came to Mazanderan for Naw Ruz (New Year) in nearly every one of the last years of his life and reign, and seemed to have been torn between the rival attractions of Ashraf and Farahabad. According to one source he passed away at Farahabad, while other sources state that it was at Ashraf: agreement is on the date of January 1629.

Continuing to the east, the causeway of Shah Abbas led to Ashraf, now called Beshahr. This garden complex lies at the foot of wooded mountains, some five miles from the Caspian, and enjoys a fine view over the bay of Astarabad; there were only scattered farm houses in the area until Shah Abbas, in 1612, gave the order for work to start. The date was commemorated in the chronogram of Dowlat-i-Ashraf, or Dominion of Ashraf, the proper name for the site, which is usually known as the Bagh-i-Shah.

The sketch map (Plate 40) illustrates the size and complexity of the original plan. The identification of the various gardens making up the entire dominion seems quite certain, although a few minor points have not been resolved. These have to do with the extent and form of the Bagh-i-Shimal, or Northern Garden; whether the Bagh-i-Tepeh, or Garden of the Mound, was intended as a private garden for the royal ladies or as a site for the baths; and whether the Bagh-i-Haram, or Garden of the Women's Quarters, bore that name at all periods or was commonly called the anderun (women's quarters), and attached to the Bagh-i-Khalvat, or Private Garden. However, consulting the sketch plan of the area and the detailed plan (Plate 41) showing part of the area as it exists today, we may proceed to tour the site.

The main portal supported a gallery called a Naqqara Khaneh, or Drum Tower, where, as at Isfahan and other large towns, a group of brass players and drummers ushered in the dawn and played the sun to rest. As early as the eighteenth century, and perhaps earlier, one could see over the entrance the royal insignia of Iran, the lion with a rising sun behind it. Passing through the portal one came into the Bagh-i-Shimal, where the guards had their living quarters and where a few Persian rugs were strewn around a white marble pool. At this point ambassadors and dignitaries were met by ranking court officials and entertained until they were summoned to the royal presence.

Many of these audiences were held in the next garden to the south, the Bagh-i-Chehel Sutun, or Garden of the Forty Columns. The way led along a central avenue and up a series of terraces until the principal building was reached: it stood behind a pool some thirty to forty meters square and of considerable depth. No description of this building has survived nor do we know its name, only that it burned down and was replaced during the reign of Nadir Shah. It was this newer structure that was called the Forty Columns and that gave its name to the older garden. That the building was altogether charming may be seen from an engraving made in 1847 by the artist accompanying Hommaire de Hell (Plate 42). Only twelve columns held up the lofty porch but it is well known that the Persians used the number forty as a synonym for a large number. Details concerning this particular garden come from a number of travelers. One speaks of cypresses more than sixty feet high and groves of orange trees more than twenty-five feet in height, while another describes the manner in which night receptions were arranged. Holes cut in the stone linings of the central water channels took as many as one thousand lighted candles, while lights around the great pool gave it the name of Pool of Lights.

In the area just to the west of the Bagh-i-Chehel Sutun were to be found the Bagh-i-Haram and the Bagh-i-Khalvat. The latter contained the personal living quarters of the ruler and was two stories in height. It was adjoined directly to the north by the women's quarters, where a very extensive structure surrounded an open court containing a large pool with marble benches in each corner angle. These two groups were tied together by a north-south water axis which continued on south to the Bagh-i-Sahib Zaman, or Garden of the Lord of the Age. At the entrance to the southernmost garden stood a lofty gateway crowned by a sheltered balcony, and upon occasion the shah would mount into the balcony while his visitors and courtiers took their places below on stone benches or on carpets spread beside a pool.

The Bagh-i-Sahib Zaman contained a single building which appears to have been used exclusively for royal functions and for storing and displaying royal possessions. The evidence of the ruins indicates that it was two stories high, that it had at least six very large chambers or halls, and that its flat roof served as a place of reception.

Although Thomas Herbert does not name the building, it seems certain that it was in this structure that he was received by Shah Abbas. To follow his own words: "Thence . . . to another Summer House, rich in gold inbosments and painting but farre more excellent in a free and royall prospect; for from the Tarrasses were viewed the Caspian Sea. . . . The ground chambers were large, quadrangular, archt, and richly guilded above and on their sides; below spread with the most valuable Carpets of silk and gold; in the center were Tancks full of crystallin water; round about the Tancks were placed Goblets, Flagons, Cesternes, and other Standards of pure massie gold, some of which were filled with Perfumes, some with Rosewater, with Wine some, and others with choisest flowers. . . . [We] were brought into another square, large upper chamber, where the roofe was formed into an Artificall Element; many golden Planets attracted the wandering Eye to help their motion. The ground was covered with richer Carpets . . . the Tanck was larger, the material more rich in Iasper and porphyr; the silver bubbly stream was forced up into another Region. So much in gold, in vessels for use and ostentation that merchants adjudged it worth twenty millions of pounds sterling; another water Magazeen there was circled with a wall of gaold and richest Ieams: no cups, flagons, nor other there, but what were very thick and covered with Rubies, Diamonds, Pearles, Emrald, Turquises, Iancinthes, etc. The Chamber was gallery wise, the seeling garnished with Poetique fancies, gold and choicest colours . . . one Iohn a Dutchman who had long served the King celebrated his skill . . . the ground in this room overlayed with carpets . . . and the round the courtiers sat cross-legged and at the upper end, slightly elevated, sat the Pot-shaw, cross-legged."

Of course, the "Pot-shaw" was the Padshah, Shah Abbas, and we will omit his words of welcome to the embassy from England. At least two of the rooms seen by Herbert featured playing fountains, their basins made of rare marbles inlaid with precious stones, while treasures of gems and costly containers were everywhere on display. We may question whether the stage was not set to make the maximum impression upon foreigners from a land so distant that the Persians had only the vaguest of ideas about it.

Herbert's concentration on the displays of wealth caused him to neglect his customary report on the mural paintings, and we have to look to the words of Jonas Hanway, who visited the garden more than a century later, to fill the gap. He remarked: "We were conducted to a banqueting house [and] out of respect to this place were required to leave our swords at the door. The solemnity with which we were conducted struck me with a kind of religious awe; but this was soon changed into contempt; for I was surprized to find the room adorned with paintings, such as could please only a voluptuous MAHOMMEDAN. Here were also portraits of Shah Abbas the first and second, and of some other persons, all by a EUROPEAN hand, but meanly performed."

Three gardens lay to the east of the Bagh-i-Chehel Sutun. One was the Bagh-i-Zaytun, or the Garden of the Olive Trees, and this has vanished without a single one of these trees remaining. The Bagh-i-Tepeh, or Garden of the Mound, represented an unusual conceit in that this section was built up some ten meters above the rest of the area. One of the earliest accounts states that the baths were located on the mound among orange trees, but another states that the area was reserved for the ladies of the court. It is certain that it received a copious water supply in its pool; de Morgan gives a sketch of lead pipes embedded in the coping of a brick wall and states that this was the means of conveying water to the top of the mound, while Hommaire de Hell speaks of siphons. Today the stone retaining walls are well preserved: over one of these walls the excess water fell into a channel leading into the Bagh-i-Chehel Sutun.

Finally comes the Bagh-i-Chesmeh, or Garden of the Spring. In this case the spring is just a few paces to the south of the existing pavilion. Nearly every early traveler mentioned the attractive Palace of the Fountain and the garden which stepped down in terraces at regular intervals. Hanway noted that the palace was crowned by a stately dome and that the walls were covered with Dutch tiles as high as the gallery; these were enameled faïence tiles. Others remarked on the paintings of Diana and the hunt, and of European and Chinese ladies, some by John Duckmann, the same John the Dutchman of whom Sir Thomas Herbert wrote.

In this attempt to recreate the complex of gardens, emphasis has of necessity been placed upon descriptions of pavilions and pools, with garden design treated only in most general terms. It is certain that the gardens themselves drew as much enthusiasm as the buildings. Hanway spoke enthusiastically of the view over the Caspian, of the backdrop of the mountains, and of the pleasing ideas which came to him from the numerous cascades and the music of the birds. A contemporary Persian account waxed lyrical over the baths, mansions, and open pavilions, and the "orchards and gardens resembling Eden and comprising these various edifices and water reservoirs of perfect beauty, filled with pleasant and salubrious water." Water and pool were present in greater abundance and number than would have been possible on the plateau, and the existence of a far-ranging view was also a feature not common to the plateau.

Ashraf went through periods of long neglect, and then of revived interest, in the post-Safavid period. Nadir Shah was fond of the site, and a leader of the Qajar tribe, Muhammad Hasan Khan, had repairs carried out just after the middle of the eighteenth century. Then local governors carried out deliberate destruction as a part of a scorched earth policy against the Turkoman invaders until Aqa Muhammad, Qajar ruler in the last quarter of the eighteenth century, once again tried to restore some of the original elegance to the site. For the space of a hundred years the structures slowly fell into ruin, and then Reza Shah (ruled 1926–41) gave orders for the resurrection of the site. That work, carried out in the 1930's, was limited largely to cleaning and clearing and to restoring the general pattern of the gardens.

Just a kilometer west of Ashraf a bold, wooded promontory represents the site of the garden palace of Safiabad, originally constructed under Shah Abbas along the plan of the Jahan Numa, and enlarged by his grandson and successor, Shah Safi I (reigned 1629–42). Early in the nineteenth century the palace and the aqueduct, which brought water from springs higher in the mountains, were repaired and a final restoration, which included the addition of modern conveniences, took place under Reza Shah (Plate 43). Today a motor road winds up the hillside and to the extensive terrace before the palace. No traces of the original gardens remain. From the terrace there is a magnificent view over Ashraf and the Caspian below; the same view was enjoyed by Shah Abbas when he came here to hunt and to drink Shiraz wine.

A number of other garden sites of Safavid and Qajar periods could be described, but mention will be made of only one more which may represent a considerable company of vanished gardens. H. L. Rabino speaks of the site of Abbasabad, just ten kilometers from Ashraf, where a palace of Shah Abbas was situated on the edge of a large artificial lake. Herbert saw the site nearly three hundred years ago, just after completion, and noted that the garden "surpasses for a curious summer house, excels all his other houses for a delicate view, Imagery, Water-works, and a Forest stored with game of all sorts."

Just as some of the Qajar rulers were sufficiently attracted by palaces built by the Safavids to undertake their restoration, so in modern times Reza Shah did much to enhance the established splendors of the Caspian. Originating from a family long resident on these wooded slopes, he came often from Tehran to visit the old sites made habitable again and to enjoy the results of his own royal orders.

Among the latter sites the most noteworthy is to be found at Ramsar, in the Gilan area of the Caspian coast. Impressed as are all visitors by the splendid backdrop of dense forest and the vista over the boundless Caspian, one may not realize that the location was selected because of its five hot mineral-water springs gushing forth sulphurous, chloride, and calcareous waters, and favored for the treatment of rheumatism, arthritis, and sundry similar ills. A sanatorium with main building and separate residential buildings marks this site, and not far away is a royal palace and the pleasant hotel, featuring plentiful private baths and an atmosphere of peace and calm. Groves of orange and lemon trees surround the buildings, and from the main entrance of the hotel an extended axis of more than a kilometer leads down to the shore. This entire distance is treated as one continuous rose bed for on either side of the paved roadway are beds of roses displaying every possible variety: walking down for a swim one passes thousands of bushes and the air is heavy with perfume. At the same time there is nothing typically Persian about this long, narrow garden, and greater enjoyment comes from either the mild winter climate or from the summertime sea breezes.

## CHAPTER FIVE

# The Royal Gardens at Tehran

For centuries Tehran had a placid existence as a modest village noted only for its excellent fruit and its beautiful, veiled women. Overshadowed by the fame of Rayy, just to the south, it lay adjacent to a main east-west route across the country. Slumbering on a gently sloping plain, the Elburz Range immediately north sheltered the site, and its mountain streams offered the potential for growth.

When Aqa Muhammad Khan, founder of the Qajar dynasty, extended his hold over the country, he originally chose for his capital Sari, on the Caspian coast near the territory of the Qajar tribe. However, in 1788 he moved the capital to Tehran, selecting a site which gave easier control over the region of the plateau proper. During his reign the population of the town remained at about fifteen thousand people, and comparatively little was done to enhance the area. When his nephew Fath Ali Shah, who ascended the throne in 1797 and ruled for nearly forty years, came to power, the situation changed. Gardens, palaces, public squares, government buildings, and private houses were constructed. Merely to meet the demands of the court, considerable construction had to be rushed through, for the monarch himself was the progenitor of some two thousand children and grandchildren. Work was begun on the Gulistan Palace at this time and the Qasr-i-Qajar (Castle of the Qajars) and Negaristan (Picture Gallery) were entirely erected during his reign.

Under his grandson and successor Muhammad Shah (ruled 1834–48) the pace of development fell off, but during the long reign of Nasir ad-din Shah (1848–96) the town became a teeming, prosperous city, assuming the form and plan it was to retain until the second quarter of the present century.

Since 1925 the character of the city has altered in many ways. The population has grown to more than one and a half million people. The southern section is much less changed but wide avenues have been slashed through the crowded quarters and narrow lanes. The town itself creeps steadily to the north: the Negaristan, which lay a kilometer outside the town in the nineteenth century, was overrun by 1930, and the Qasr-i-Qajar, four kilometers to its north, was overtaken between 1930 and 1955.

Tehran's noted suburbs are those bordering the abrupt and steep slopes of the Elburz Range, some eight kilometers north of the city. Known collectively as Shemiranat, the region is dotted with the villages of Shemiran, Gulhak, and Tajrish. Two wide boulevards connect with the capital, and a third is in the planning stage. In earlier years the spring exodus up into these slopes moved the establishments of the royal court, the nobles, and the diplomatic colonies so that they could enjoy an abundance of water and shade and a night temperature that was fifteen or twenty degrees cooler than that of the town. A handful of royal palaces and gardens were dotted along the slopes, together with many private gardens and the residences of the Russian, German, and other diplomatic missions. However, the character of this region also has altered a great deal in more recent years. More and more houses go up as more and more Tehranis live the year round in the Shemiranat, inconvenienced only by occasional heavy snowfalls in the winter. Many of the larger gardens have been cut up, and many of the spacious, high-ceilinged pavilions torn down. The government has taken steps to preserve and maintain such sites as the royal palace of Sahib Qaraniya at Niavaran, and the palace site of Sa'adabad, developed by the present dynasty, stretches over many acres.

Every visitor to Tehran who sees any of the sights, manages to take in the Gulistan Palace with its two gem-studded thrones and its museum of nineteenth-century European treasures. To illustrate an account of this area, paintings made about 1865 by three of the noted Persian court artists of the time are available.

The Gulistan Palace was erected during the reign of Fath Ali Shah and appears to have been completed by 1806 when the ruler received a Frenchman, P. A. Jaubert, who later described this experience. Jaubert entered the area through the Dar-i-Sa'adat, or Gate of Happiness, also known as the Ali Qapu, or High Gate. The painting by Mirza Abdul Hasan Khan Sani'a ol-Molk (Plate 44) shows us this portal and one of the vast pools so typical of the Qajar period at Tehran. Within the portal one looks directly ahead, along the axis of a long, fairly narrow pool, to the structure called the Divan Khaneh, or Reception Hall, or sometimes the Takht-i-Marmar, or Marble Throne (Plate 47). We have an account of a dazzling reception held here; ten pages would not contain the details of Robert Ker Porter's 1821 account. Along the edge of the pool were ranks of plates, each heaped high with oranges, pears, apples, grapes, and dried fruits, and between the plates were tall vases filled with wax flowers. Beyond were regular rows of bowls of the finest china filled with sherbet.

The structure is recognizable as the typical, southern-oriented pavilion, but magnificent in every way. The finest rugs covered the floor of the open porch, the walls were of mosaic work, the ceiling done in mirror work and with pictures of the hunt, of battle, and of royalty in every niche. Dominating all was the white marble bed-throne which still stands in the same position. As the Shah took his place he seemed to be "one blaze of jewels, which literally dazzled the sight on the first looking at him, but the details of his dresses were these: A lofty tiara of three elevations was on his head.... It was entirely composed of thickly set diamonds, pearls, rubies, and emeralds, so exquisitely disposed, as to form a mixture of the most beautiful colours, in the brilliant light reflected from its surface. Several black feathers were intermixed with the resplendent aigrettes of the truly imperial diadem. . . . His vesture was of gold tissue, nearly covered with a similar disposition of jewelry and, crossing the shoulders, were two strings of pearls, probably the largest in the world. . . . The jewelled band on the right arm was called The Mountain of Light (Kuh-i-Nur); and that on the left, The Sea of Light (Darya Nur); and which superb diamonds Nadr Shah had placed in the Persian regalia, after sacking Delhi . . . and [of] the treasure transported thence to Persia . . . no part of it was so highly prized as these transcendent precious stones. . . . [As the Shah] approached his throne . . . the whole assembly continued bowing their heads to the ground. . . . A dead silence then ensued . . . the stillness being so profound, amongst so vast a concourse, that the slightest rustling of the trees was heard, and the softest trickling of the water from the fountains into the canals. [Suddenly] a volley of words [burst forth]. This strange outcry, was a kind of heraldic enumeration of the Great King's titles, dominions, and glorious acts; with an appropriate panegyric on his courage, liberality, and extended power." Not only was the reception a unique experience for Ker Porter but it paid off as well, for after he had enjoyed a delicious sherbet an attendant presented to him a large silver tray on which lay a heap of small silver coins mixed with a few of gold, and invited him to take two handfuls!

From this area the way leads east to the largest open area of the gardens, with important structures on the north and east, and lesser ones along the other sides. To the north was a structure shown in a painting done in the mid-nineteenth century (Plate 45). This structure, now replaced, contained an audience chamber which opened to the exterior on two opposite sides, and the illustration shows a bed-throne on the edge of the pool. By 1875 this side of the palace had been rebuilt as it appears in the background of another painting (Plate 46) and as it remains today. The arched portal gives access to a stately flight of stairs and to a gallery overlooking the garden. This gallery opens into the great hall of audience. Many displays of pomp and circumstance have taken place in this hall. Reza Shah was crowned here, and each year on Naw Ruz (New Year) the reigning monarch receives his officials and foreign dignitaries in audience. At the far end of the hall are the two thrones about which so much has been written. Whether or not the larger is really the Peacock Throne taken from Delhi, and whether or not original gems have been replaced by glass, the thrones do make a splendid display.

Alcoves along the walls and in bays hold a great variety of trash and treasure. Here were assembled gifts from the European diplomatic missions and results of the nineteenth-century Persian shopping tours of the continent—purchases made by traveling shahs and items sent to fill standing orders for the unusual and exotic. The initial installation of this museum was by Nasir ad-din Shah in 1872. Delicate mechanical clocks stand near early-model alarm clocks, and pictures in Italian stone mosaic are alongside scenes made from inlaid wood or from butterfly wings.

The towering structure on the east was the Shams ol-Amareh, or Building of the Sun, shown in a painting of 1868 (Plate 49). The pool has been replaced by planting, and the main floor of the building now houses the marvelous collection of illustrated manuscripts belonging to the crown.

The garden itself covered an area of more than an acre. According to one early account it was bright with tulips, narcissi, anemones, and poppies, sown as if by chance over a green lawn. At this same time, in 1806, a tent was pitched at the very edge of one of the great pools. Rich silk drop curtains could be lowered for shade or privacy, and on summer nights Fath Ali Shah and his selected partner of the evening would pass the hours in the garden, just as if they were out in the open country.

In the section of the town near the Gulistan Palace and the bazaar were a number of fine palaces and gardens. Typical of these structures is that which housed a French mission of the mid-nineteenth century and was recorded by the mission's artist (Plate 48). The view is south; the pool is a great feature, and the light fences seem scarcely strong enough to contain so much foliage.

The favorite residence of Fath Ali Shah was the Negaristan, which was situated a short distance to the northeast of the town. Some years ago the area was taken over by the Ministry of Education and modern buildings replaced the royal structures.

The Negaristan was completed about 1810; its pavilions were set within an oblong enclosure of several acres. The entrance portal was on the long axis and decorated on the exterior with enameled tiles. The central avenue led to a Kola Ferangi which was situated at the very center of the garden and approached "between the spacious arcade of trees. The trees were all full grown, and luxuriant in foliage; while their lofty stems, nearly covered by a rich underwood of roses, lilacs, and other fragrant and aromatic shrubs, formed the finest natural tapestry of leaves and flowers." The most striking feature of this Kola Ferangi was the large pool and playing fountain set in the center of the main room.

At the far end of the axial avenue lay the most important structure, variously known as the Negaristan or the Divan Khaneh. Ker Porter wrote that it was nearly circular; it was probably an octagon. Ker Porter's account includes charming detail: "At the upper end of the garden, is a small and fantastically built palace, enclosed in a little paradise of sweets. The Shah often retires thither, for days together, at the beginning of summer . . . and accompanied by the softer sex of his family, forgets, for awhile, that life or the world have other seasons than the gay and lovely spring. The building was of a light architecture, and, with its secluded garden, presented altogether a scene more congenial to the ideas I had conceived of one of these earthly imitations of the Houris' abodes, than any I had yet met in the East. The palace was nearly circular, full of elegant apartments, brilliantly adorned with gilding, arabesques, looking-glasses, and flowers natural and painted, in every quarter. Some of the largest saloons, were additionally ornamented with pictures; portraits of the Shah and his sons; of the chief personages at court; also of foreign ministers; and amongst the rest, were General Sir John Malcolm, Sir Hartford Jones, Sir Gore Ousely, Monsieur Gardanne, etc., etc., etc., all pourtrayed in high costume, and all like one and the same original. . . ." Famed for a number of decades were these large-scale paintings: Browne noted that they were completed in 1813 by the painter Abdullah and included some 118 full-length figures. Seated on his throne and flanked by his sons, Fath Ali Shah received the British and French envoys, and everyone was dressed as elegantly as possible.

However, the pavilion which caught the eye of the European envoys and writers was the summer bath known as the Taj-i-Dowlat, or Crown of the Kingdom. This structure was located near the left wall of the enclosure, and was at least two stories in height, with the upper floor taken up by women's apartments. Below was the marble bath, with a long marble slide serving as a rapid and gay means of descent from the apartments. The visitors enjoyed visualizing the Shah catching his unclad favorites as they came down to splash in the pool; or of the ruler gazing down upon them from the upper floor.

The words of Ker Porter present a vivid picture of this atmosphere: "This bath-saloon, or court, is circular, with a vast basin in its centre, of pure white marble, of the same shape, and about sixty or seventy feet in diameter. This is filled with the clearest water, sparkling in the sun, for its only canopy is the vault of heaven; but rose-trees, with other pendant shrubs bearing flowers, cluster near it; and, at times, their waving branches threw a beautifully quivering shade over the excessive brightness of the water. Round the sides of the court, are two ranges, one above the other, of little chambers, looking towards the bath, and furnished with every refinement of the harem. These are for the accommodation of the ladies who accompany the Shah. They undress or repose in these, before or after the delight of bathing; for so fond are they of this luxury, they remain in the water for hours; and sometimes, when the heat is very relaxing, come out more dead than alive. But in this delightful recess, the waters flow through the basin by a constant spring; thus renewing the body's vigour, by their bracing coolness; and enchantingly refreshing the air, which the sun's influence, and the thousand flowers breathing around, might otherwise render oppressive with their incense. The royal master of this *Horti Adonidis*, frequently takes his noon-day repose in one of the upper chambers which encircle the saloon of the bath; and, if he be inclined, he has only to turn his eyes to the scene below, to see the loveliest objects of his tenderness, sporting like naiads amidst the crystal stream and glowing with all the bloom and brilliancy which belong to Asiatic youth."

Several visitors exclaimed with pleasure at the rose trees which were a special feature of the Negaristan. Their name, *nastaran*, seems to mean sweetbriar rose tree: the trees grew to a height of twenty feet with a trunk, of two feet in circumference. The flowers, like the English hedge rose but larger, were so abundant as to completely conceal the trunk. From these trees came a most exquisite perfume. In fact, these rose trees inspired Ker Porter to say: "In no country of the world does the rose grow in such perfection as in Persia; in no country is it so cultivated, and prized by the natives. Their gardens and courts are crowded with its plants, their rooms ornamented with vases, filled with its gathered bunches and every bath strewed with the full-blown flowers, plucked from their ever-replenished stems. Even the humblest individual, who pays a piece of copper money for a few whiffs of a kalioun (pipe), feels a double enjoyment when he finds it stuck with a bud from his dear native tree."

Multitudes of nightingales haunted the shade of the Negaristan and warbled throughout the blooming season until, as the poet says: "When the roses fade, when the charmes of the bower are passed away, the fond tale of the nightingale no longer animates the scene."

Nearly six kilometers due east of Tehran is a royal site adorned by a palace completed as late as 1904: the site has been variously known as the Qasr Farahabad, or Castle of the Abode of Joy; as Oshnan Tepeh, or Hyssop Hill; and as Dowshan Tepeh, or Hare's Hill, its current name.

The original complex was built by order of Muzaffar ad-din Shah (ruled 1896–1907). A barren hill is crowned by a glaring white palace (Plate 50) of semicircular plan; several tiers of open verandas stretch along the arc and give a most un-Persian look to the structure. According to one report, the structure was modeled after the Trocadero Palace at Paris. Each floor has one large central room and a number of oddly shaped flanking ones: these rooms hold the debris of a royal collector, things that are not worth stealing. These dilapidated treasures include a few fine small rugs, stuffed birds, animal skins, indifferent oil paintings in all stock sizes, and an amount of bronze statuary. However, the structure itself may be of less interest than its setting.

As one approaches the palace, the road passes a small hill which was once the site of a royal menagerie. Here were specimens of native lions, tigers, and leopards and of imported animals, including monkeys. In its final years, the enclosing walls fell into ruin and the beasts were no longer fed; some escaped and others devoured the weaker ones.

Three separate large enclosures comprised the palace and garden area.

The first held the service quarters, including an oval brick stable, and the second the orchards. The third area is still maintained as a garden and features a very large pool on the axis of the enclosure. However, the garden area is developed asymmetrically with attention centered on the eastern half: the treatment is so Europeanized that little of the characteristic Persian atmosphere remains. There are radiating alleys, unbalanced features, and a charming group of structures in wrought iron: a greenhouse, a many-sided tea pavilion, and a grape arbor (Plate 51). This site is sufficient evidence to indicate the decay of Persian garden art before the onslaught of the West: by 1900 it was impossible to be "modern" without copying from England, France, or Russia.

Just north and east of Tehran lay the royal garden called Eshratabad, or Abode of Pleasure. Accounts agree that it featured one lofty structure called Khabgah, or Sleeping Palace, and seventeen small buildings for the royal favorites. In 1880 the painter Mahmud Khan Malik esh-Sho'ara recorded a part of the complex showing the buildings around a vast, circular pool, but gave no impression of the garden itself (Plate 52).

West of the town was the site of the Bagh-i-Shah (Royal Garden), shown in a painting made in 1888 by Muhammad Chaffari Kamal ol-Molk (Plate 53). In the center of an artificial lake was an island reached by a bridge and featuring an equestrian statue in bronze of Nasir ad-din Shah. As at Dowshan Tepeh the details are all Europeanized, from the low metal railings to the extreme formality of the flower beds, and our Persian painter may have found this design so inharmonious that his attention wandered and his landscape sags downhill at the left of the picture. Lake and statue have vanished but the site still bears this name.

Upon its completion by Fath Ali Shah the Qasr-i-Qajar, or Castle of the Qajars, rivaled in splendor and surpassed in size the other gardens of the period, yet less than half a century later it lay abandoned, housing colonies of swallows instead of bevies of beauties. Situated some four kilometers north of Tehran and just to the right of the first motor highway to Shemiran, the ruins have been seen by thousands of foreigners, few of whom were aware of the significance of the site.

As the restored plan shows (Plate 58), the original layout comprised an extremely spacious enclosure on level ground and a walled-in structure which rose in terraces up the rather steep slope. At the foot of the slope was a great pool (Plate 54) designed in the tradition of combined hillside garden and artificial lake as at the Shah Gol, near Tabriz, and the Bagh-i-Takht at Shiraz. In the 1930's the area of the enclosure was taken over as the site of a model prison for Tehran, while the upper building endured into the 1950's, when it was robbed for brick and stone for a nearby housing development. In describing the garden as it appeared more than a century ago, we follow the plan which has been drawn up from several accounts of the site.

This plan appears to reflect faithfully the original arrangement of the twenty-acre enclosure, since an early witness states that it was "laid out in parallel walks, planted with luxuriant poplars, willows, and fruit-trees of various kinds, besides rose-trees in profusion." At the center of the garden stood an imposing kiosk erected of green marble and brick and clad with enameled tiles (Plate 55).

At the northern end a guarded portal gave access to the castle proper, while above the portal was a spacious gallery decorated with paintings. The terraces were traversed in all directions by streams of water, and the fifth terrace led to the private residence of the ruler. Rooms surrounded the open court and much of the court area was filled by a pleasant pool. The rooms at the upper end housed the Shah himself, in particular one chamber that was inlaid with ebony and ivory and mosaic work and decorated with paintings on tiles. From this vantage point the view ranged far over Tehran, the site of ancient Rayy, and the shrine of Shah Abdul Azim with its gleaming dome: it remains a moving sight to gaze across this plain just before sunset when the sky is gold and red, when long fingers of shadow steal for miles across the earth, and when countless columns of smoke rise straight into the clear air. The other rooms were for the ladies of the court; each was only about four meters square and almost filled by a huge wooden bedstead. Paintings in superior taste and skill covered many wall surfaces: some depicted Persian rulers and heroes of history and fable, some women in European dress, and others members of European missions to Iran. Notable in the last class was a striking full-length figure of an Englishman, a Mr. Strachey, called by the Persians, Istarji, who made such an immediate impression on Fath Ali Shah that the ruler himself penned an ode to him. This poetry has outlived the picture of Istarji in his court dress, with knee breeches and sword.

The vanished royal garden at Shahrestanak featured a structure called, in all simplicity, the Emarat, or Edifice, which was recorded by two painters of the period. One painting showed a good part of the entire enclosure (Plate 56) while the other, painted in 1880 by Mirza Mahmud

Khan Malik esh-Sho'ara, concentrated attention upon the building itself. Flags fly in a stiff breeze, gardeners are at their business, and all await the arrival of the monarch. The rocks in the background are not hurtling through the air, but reposing on the hill behind the edifice; the hill shows clearly in the other painting. Shahrestanak is some forty kilometers north of Tehran at an altitude of nine thousand feet.

During the nineteenth century a number of royal gardens were brought into being in the Shemiranat, that fertile strip along the base of the abrupt slopes of the Elburz Mountains some eight kilometers north of Tehran. There, in localities with names such as Gulhak, Tajrish, Sultanabad, and Niavaran, the rulers and their nobles sought refuge from the heat of the Tehran summer in the dense shade, rushing water, and cool nights of the mountains. There are too many sites to permit even a brief statement about each one, so some must be eliminated from consideration. For example, Sultanabad, a handsome development of Nasir ad-din Shah, is inaccessible, as it is now part of a military establishment. It is also proper to bypass the ruined Bagh-i-Firdaws, or Garden of Paradise, and the Qamraniya, or Place of the Two Moons.

Well preserved and maintained is the palace of Sahib Qaraniya at Niavaran. This name surpasses in elegance all those other stately, proud names recorded thus far, for it means "the place of one born at the time of the conjunction of two happy stars." Most of the existing development dates from the middle of the nineteenth century and the reign of Nasir ad-din Shah. The main structure parallels the mountains and offers a superb view over Tehran. The gardens on the south, or Tehran, side of the building are run down, but the area to the north is well looked after by a staff from the Ministry of Court. This area is sprinkled, as was Eshratabad, with a number of self-contained buildings, each just large enough to house a lady friend and her servants; originally there were forty such residences.

The garden was endowed with its own qanat so that an abundant flood of water poured into the tile-lined channels which ran through every section of the enclosure. From the wooded area at the north the channels emerge into a cleared area just in front of the main building where there are elements of a formal garden, combined with stages and steps for the display of potted plants. The formal beds displaying patterns in blossoms and greenery are foreign inspired, but the ranks of potted plants are entirely in the Persian tradition.

The building itself is in a style of architecture imported from Russia, and the central open porch facing south is gone, although in this case the colder climate may have militated against its use. A huge reception room on the second floor looks north to the mountains and south over the plain. This room is furnished in the height of Persian nineteenth-century luxury and taste, with objects imported from France and Russia especially for this building. There are elaborate crystal chandeliers, man-sized porcelain vases, and—most unusual—a grand piano with two keyboards, one at either end. This monumental object had to be transported from the port on the Caspian over the eight-thousand-foot pass of the Elburz Range, and a distance of one hundred miles, slung from the backs of camels. How much material must have been lost in transport, particularly such items as the huge panes of plate glass that are featured in some of the Qajar buildings! However, there are Persian touches in the building, including fine carpets, brocades, and embroideries. A corridor wall is hung with the paintings of Mahmud Malik esh-Sho'ara; his paintings of the Gulistan illustrate this chapter and it is regretted that he did not record the Sahib Qaraniya.

The charm and attractiveness of the Shemiranat is as lively as ever, and scores of modern gardens dot the region. Of this number only two will be mentioned. One continues the royal tradition and the other displays private elegance. The private garden, belonging to the Firuz family and featuring a wide variety of beautifully cultivated flowers, is called Bagh-i-Chal, or Garden of the Pit. Entering through an inconspicuous door in a mud-brick wall, one comes directly on the beds where dahlias, cannas, begonias, and gladioli grow to mammoth size in separate cutting beds. Beyond is a spacious open area where the traditional hillside garden and lake garden have been executed with modern touches. The expanses of cut stone, the metal railings, and the beds of fiery red cannas are departures from the older norm, but the entire effect of playing fountains and contrasting colors of foliage and flowers is harmonious and stimulating.

The royal gardens of Sa'adabad cover more than forty acres in a section north of the maidan at Shemiran. In May or June the Shah and his Minister of Court move up from Tehran for the summer. The garden area displays a variety of treatment and encompasses nearly a score of buildings. In addition to the building housing the Ministry of Court, there is a White Palace where the Shah has his own working offices and where visitors are received. In this same area is the private residential palace with its swimming pool and tennis courts. Formal gardens surround these palaces and it may be regretted that European design has taken over (Plate 57). Watercourses run through the area and a considerable part of it is wooded, offering secluded trails for riding. At one end is what can be appropriately called a monticule: on its crest perches a diminutive palace erected for Reza Shah, father of the present ruler. Here the former ruler was able to find peace and solitude in austere but princely surroundings. Some of the rooms were entirely lined with the splendid inlaid mosaic work for which Shiraz has long been renowned. The ruler's private library has been preserved; near his bedside was a copy of the translation from English into Persian of *Hajji Baba of Isfahan*, by Morier. Nearly everyone who has traveled in Iran, or taken any interest in this country, has read this delightful satire on the foibles of the Persian. To Reza Shah it may have been of surpassing interest in connection with his efforts to reconcile Persian thought and behavior with the world of the twentieth century.

## CHAPTER SIX

# Shiraz: Home of Gardens and Poets

Visitors to Shiraz in the period from the seventeenth through the nineteenth centuries concentrated their attention upon the tombs of the renowned poets, Sa'di and Hafiz, upon the gardens of the town, and upon its red wine. Few were disappointed in the wine, but some skeptics wrote that the gardens and the tombs had been overrated.

Chardin, the Frenchman who wrote such a long, detailed, and fascinating account of Isfahan, visited Shiraz in 1674 and was pleased with what he saw. He noted that Shiraz was full of gardens. He came through the Allahu Akbar Pass on the north of the town to descend along a most beautiful wide avenue lined on both sides with symmetrically opposed gardens, each marked by a portal crowned by a semidome and with a pavilion perched over each portal. Chardin wrote: "The most beautiful things at Shiraz are the public gardens, twenty in number, which contain the largest trees of their kind in the world . . . cypresses, plane trees . . . and pines." Then, going into detail, he mentions a royal garden and a Bagh-i-Firdaws, or Garden of Paradise, which featured a pool that was 125 paces along each side. He describes the red wine in some detail: its preparation, its taste, and its export to India, China, and Japan in glass bottles sealed with cotton and wax. The handblown bottles of Shiraz are still sealed in this way. While Chardin does not compare the taste of the wine with European types, this may be a proper place to say that it resembles a Malaga.

Edward Scott Waring, who came to Shiraz in 1802, gives a somewhat less enthusiastic account, stating that he saw little at Shiraz worthy of the lavish praises of its inhabitants. However, his words serve to put the picture in correct perspective, as he quite rightly states that the gardens were meant to charm the inhabitants and not the foreigners. He wrote: "The gardens about Sheeraz are much celebrated; but the striking uniformity of long walks and narrow alleys is sure to displease European taste. You may, perhaps, walk a quarter of a mile, and on either side not have a view of a few yards. Yet the Persians delight in visiting these gardens; anything delights them; and a running stream almost makes them frantic. Nor is this to be wondered at; it is here that they rejoice themselves from the anxiety and drudgery of business, and enjoy their Sohbuts (sherbets). The day is passed in smoking, in the amusement of fishing, or in listening to the odes of various poets. . . ."

These travelers have plunged us into the midst of the gardens, and we should pause to approach them in a more leisurely fashion. What of the history of gardens at Shiraz? Early references are few and scarcely revealing. A local ruler at the end of the tenth century built himself a palace with many fine gardens; and a *Shiraz nama*, or History of Shiraz, composed in 1335, mentions several gardens but without giving any details. One was in combination with the Masjid-i-Naw, the New Mosque, which still stands at the center of Shiraz. However, few of the existing gardens are known to be older than the eighteenth century.

The plan of Shiraz included here (Plate 59) shows the location of the important gardens existing today and also suggests the positions of some which have disappeared. As the traveler approaches Shiraz on the road from Isfahan, the last stretch is along a winding road among stony hills. At last Shiraz bursts into view at the pass called Allahu Akbar (God is Great), for so travelers ejaculate at their first sight of the rich verdure, colored domes, and mountainous backdrop of the famous town. Just down the grade from the pass, the road once proceeded through an entrance gateway called the Qoran Gate (Plate 60), this name deriving from a Qoran of huge size which was formerly kept in a room directly over the arch spanning the road. A number of years ago this gateway was torn down in order to facilitate motor traffic, but in 1948 a local philanthropist, Husayn Igar, erected a new structure clad with faïence and with a central passageway wide enough for two trucks to pass abreast.

To the left of the avenue, which leads down a slope, across the scanty Kushk River, and into the heart of the town, flows the rivulet once famed as the stream of Ruknabad. Formerly it flowed through a pleasure ground called Musalla before it reached the river. Today the avenue is lined with modern structures, including a group of oil-storage tanks, which make a mighty effort to ruin the view. By fitting together several descriptions of this avenue as it appeared at the end of the seventeenth century, we are able to discover that it was of the same type as the Chahar Bagh at Isfahan. A canal of water bordered by cypress trees paralleled the avenue and at two points filled octagonal basins.

Hafiz sang the praises of this northern outskirt of Shiraz in these words (from the translation by Gertrude L. Bell):

> Bring, Cup-bearer, all that is left of thy wine!
> In the Garden of Paradise vainly thou'lt seek
> The lip of the fountain of Ruknabad
> And the bowers of Musalla where roses twine.

Hafiz (Shams ad-din Muhammad Hafiz), one of the greatest poets of Iran, and of the world as well, was born at Shiraz on a date which is not precisely known, and he died there in 1389. He displayed real devotion to his birthplace and many of his verses tell of its charms, as these lines:

> Joy be to Shiraz and its unrivaled borders
> O heaven preserve it from decay.

His collected works, which include nearly six hundred odes, are familiar to nearly all Persians, and he stands highest of all Iranian writers in popular esteem. However, the life and the works of Hafiz are not to be discussed here, but only his garden-embellished tomb.

Hafiz was laid to rest in his beloved Musalla, and in time the site came to bear his own name, the Hafiziyya, or Place of Hafiz. History records that the site was beautified in the fifteenth century, and early in the eighteenth century the local ruler, Karim Khan Zand, had the area put in order and contributed an alabaster tombstone, inscribed with some of the poet's own words. Later on, a structure of iron grille work was built around the slab but in recent years a graceful, octagonal kiosk (Color plate 11) has taken its place, and a colonnade has been built across one side of the open enclosure in which the tomb is found.

For many years the Hafiziyya has been a favorite resort of the Shirazis, while the poet's grave is surrounded by the closely packed graves of many others who were eager to be associated with the illustrious remains. Visitors drink tea, recline in the shade of the spreading pines, and chant the poet's odes. Many pilgrims come to the spot to make a *tafa'ol*, or augury. The *Diwan*, or Collected Works, is opened at random and one reads either the first verse which meets the eye or the last ode on the open page: these verses will answer any question that was in the mind of the reader, though some of the answers may be a bit vague.

Sa'di, another son of Shiraz, enjoys almost as much renown as does Hafiz. Born about 1181 at Shiraz, he is said to have died in 1291. Unlike Hafiz, who stayed close to home, he traveled in India, Arabia, Africa, and Asia Minor. His chief works are known as the *Gulistan*, or Rose Garden, and the *Bostan*, or Orchard, so that he is a most proper subject for our consideration. The first of these works is a series of anecdotes written in prose, interspersed with passages of poetry, while the second is entirely in verse and emphasizes the cultivation of virtues. Himself a man of moderation, Sa'di preached moderation and concern for upright behavior in this life as proper preparation for the life to come.

Karim Khan Zand is said to have restored the site of Sa'di's tomb, the Sa'diyya, located about a mile to the northeast of the Hafiziyya at the head of a narrow valley, where for many years visitors were shown a manuscript of the complete works of the poet dating from the time of that ruler. Most of these same visitors spoke of a pool inhabited by fish of considerable size which were considered safe from marauders since they were under the poet's protection. The charm of the Sa'diyya before its recent renovation consisted largely of the pines and cypress, the dramatic location at the entrance to the valley, and the deep, sizeable spring in a cleft in the ground. However, the tombstone of the poet was enshrined within an unpretentious structure of the nineteenth century. The residents of modern Shiraz decided it was inadequate for the purpose, and in 1952 the Shah of Iran dedicated a new building, erected after the designs of a Persian architect. The style of the structure might be described as elongated neo-classic in that it has a contemporary appearance, while retaining such local features as the use of faïence mosaic decoration and a blue faïence-coated dome. The park-like, formalized aspect of the renovation illustrates the Westernization of the concept from that of the traditional intimacy of a Persian garden.

From the tombs of the poets it is rewarding to move on to consider other gardens in this same general area, those located on either side of the previously described avenue leading down from the Allahu Akbar Pass into the town. Quite some distance to the east of the avenue is the Bagh-i-Dilgusha, or Garden of Heart's Ease, located in the region of the Sa'diyya. In 1812 James Morier spoke of it by this name and described it as a work of Karim Khan Zand which had fallen into a neglected condition. Histories written in Persian—one of the province of Fars and one of the ruins of this part of Iran—record that in 1737 it was owned by Taqi

Khan Shirazi and about 1845 was acquired by a member of the Qavam ol-Molk family (today known as the Qavam-Shirazi family). The garden had its own qanat, and for years its pool and fountains were much admired, while its avenues were lined with orange trees. The garden remains in the latter family, and the pavilion (Plate 61) is kept in good condition as a family residence: plaques in the interior list the names of its former owners.

Just to the north of the Hafiziyya are to be found the gardens named Haft Tan, or Seven Bodies, and Chehel Tan, or Forty Bodies, the names referring to the number of saintly personages allegedly buried within the high mud-brick walls. Both are ascribed to the time of Karim Khan (the third quarter of the eighteenth century) and this tradition may well be correct since both gardens were seen by European travelers as early as 1810. Both belong to the type of the modest garden dwelling in which a pavilion is placed at the northern end of a walled plot; the pavilion is the width of the plot, the rooms open south, and a pool is placed immediately in front of the pavilion. The Haft Tan is shown in plan (Plates 62, 63).

It is now necessary to speculate as to the locations and features of vanished gardens known to have been in this vicinity. There is ample evidence that the Bagh-i-Jahan Numa, or Garden of the Image of the World, and the Bagh-i-Naw, or New Garden, faced each other across the wide avenue already described. The former was to the east of the avenue and quite near the Hafiziyya and the Haft Tan. It is possible to trace it back as far as 1607 when Shah Abbas planted a cypress tree at this location in a garden which became famous for its stately cypresses. In 1766 Karim Khan Zand established the Bagh-i-Vakil, or Garden of the Regent, at this location but after only half a century the name was changed to the Bagh-i-Jahan Numa. Several accounts agree on the fact that it occupied an enormous square tract which was walled round about: it seems to have been artificially elevated above the surrounding area and laid out in a network of walks lined with cypresses and plane trees.

Visitors of the first quarter of the nineteenth century spoke of artificial cascades, of several marble-lined pools, and of marble water channels. Over the entrance on the avenue was a pleasure house of the gayest and most elegant construction: rooms were wainscoted with Tabriz marble and there were mural paintings on the upper walls. At the center of the garden was a pavilion, probably octagonal, called the Kola Ferangi, or European Hat. As I have said, this latter name is a generic term applied to a type of isolated belvedere or other garden structure. The main room of the pavilion featured a pool and a playing fountain. In the corners or angles of the structure was a series of mural paintings: pictures of a marriage ceremony and procession, hunting scenes, and scenes of popular entertainment including dancing bears. In its time the most lavishly decorated garden of Shiraz, the Bagh-i-Jahan Numa has vanished without a trace.

The Bagh-i-Naw is charmingly illustrated in a drawing made in 1840 (Plate 65). The early site may be identified with the location of the present Park Sa'di Hotel, although it is questionable whether traces of the early pavilion are to be found in the larger ensemble of the present building. The first embellishment of this garden is assigned to the year 1810 and to Nawab Vala Husayn Ali Mirza Farman Farma—just one man—who was the eldest son of Fath Ali Shah, the Qajar ruler of Iran. Farman Farma was governor of the province of Fars for forty years and displayed a true devotion to his second home. The site is still marked by a pool of imposing size which is supplied by a qanat, and the local stone carving on the main façade of the building (Plate 64) recalls that shown in the early lithograph and is in the same style as a number of reliefs on other structures at Shiraz. This site features a fine view of the town.

Clearly visible from the presumed site of the Bagh-i-Naw is the crumbling complex known today as the Bagh-i-Takht, or Garden of the Throne. Some hundreds of meters to the west, it abuts a rocky hillside and owes its existence to the presence of a spring which gushes from the rock. An early history of Shiraz indicates that a local ruler of the eleventh century, Atabek Qaracheh, was responsible for constructing a garden at the spot, and it is mentioned again in the seventeenth century when Herbert describes it under the name of "Hony-Shaw." It is easy enough to turn his phonetic transcription into the proper Khaneh Shah, or House of the Shah.

About the year 1665 Jean Baptiste Tavernier describes a Garden of Paradise which can be identified with this site. Tavernier writes: "Without the City, upon the North-side, at the foot of a Mountain, is a Garden belonging to the ancient kings of Persia, called Bagh-i-Firdaus. It is full of fruit trees and rose trees in abundance. At the end of the Garden upon a descent of a Hill stands a great piece of Building and below a large Pool affords it water." In 1705 Cornelis de Bruyn wrote of "the amiable palace of Ferodus," but just a hundred years later the garden had acquired the name Takht-i-Qajar, or Throne of the Qajar (dynasty). At that time its foundation was ascribed to the Qajar ruler Aqa Muhammad Qajar and to the year 1789; on the left of the area was an enclosed ground for antelope and other game. At some unknown date after 1850 it acquired its present name, the Bagh-i-Takht. A photograph taken about a half century ago shows the garden in a ruined and neglected condition (Plate 66). Since that time the structures on the upper level have been restored, and the plan depicts the site as it appeared in 1945 (Plate 67). However, the great pool now stands empty and dry. Even in ruin the garden is of real interest for it is thoroughly typical of the type of garden fed by a spring, the flow of which is collected into a large pool. As I have said, such a pool was called daryacheh, or little sea, and it is certain that it was made as large as possible. Descriptions emphasize that all such pools were provided with small boats; in these boats the proud owners navigated across the open sea, leaving their arid shores far behind. The type of hillside garden appears in Azerbaijan and then again in India where Persian influences may have been grafted onto the plans established by Babur so long before. The fact is, there is no way of knowing when this type of garden appeared in Iran.

On the north of the town but nearer to the river were—and are—other well-known gardens. One that was frequently mentioned was the Bagh-i-Rasht Bihesht, or Garden of the Envy of Paradise, which was laid out about 1824. This garden may have been in the near neighborhood of the Bagh-i-Eram, today the best known to the general visitor of the Shiraz gardens. It takes its name from a fabulous garden of Arabia, cited in the Qoran as "Iram adorned with pillars," held by the Ad tribe dwelling in the region to the south of Mecca. Upon occasion the name is misquoted as the Bagh-i-Aram, or the Garden of Tranquillity, and this name does seem appropriate to the atmosphere of the garden.

Local historical sources are not so clear on the history of this garden as on others at Shiraz. There is no agreement as to the name of the founder of the garden, although it is generally ascribed to one of the heads of the nomadic Qashqai tribe. The main structure is assigned to a local architect, Hajji Muhammad Hasan, who was without a peer in the design of garden pavilions. For at least seventy-five years the garden remained in the hands of the Il Khans, or chiefs, of the Qashqai tribe, and a photograph taken about forty years ago shows the then head of the tribe in front of the main structure (Plate 68). This same structure is the focal point of the garden, as is shown in the plan of the garden, measured and drawn in 1945 (Plate 69). At this time the garden was divided about midway by a wall, and the lower section was owned by Shahriyar Rashidpur. Since that date the upper section has passed out of Qashqai hands.

The Bagh-i-Eram owes its continuing popularity to its groves of orange trees, its long avenue of stately cypresses, and the impressive structure which has been the scene of Qashqai hospitality so freely offered to so many visitors. Every few years more of the orange trees fall victim to heavy frosts, but the cypresses are as impressive as at any time in the last fifty years. In the plan arrangement the long axis is pronounced; indeed it is along this axis that is to be found all the interest of the garden, the balance of the area being taken up by irrigated plots crowded with citrus and fruit trees (Color plate III). It is easy to imagine the earlier owners leaving the main pavilion and strolling down the water-lined avenue as far as the lower pavilion, where they rested from their exercise.

Today the main pavilion is the focal point of interest. The lower rooms are just below ground level and the central room is designed for hot-weather comfort: the water channel passes right through it—filling a pool on its way—before emptying into the great basin, and the walls and floor are lined with enameled tiles. Steps lead up to the living floor and to the passageways into the great central rooms; to the south the view is prolonged down the main axis; and to the north it opens onto the hills that rim the river valley. As in so many structures at Shiraz, glazed tile and carved stone revive the ancient heritage. The pediment area depicts a Sassanian scene in colored tile, while at the ground level slabs of limestone display somewhat distorted copies of figures from the processional reliefs at Achaemenid Persepolis. Throughout the extensive area few flowers bloom in beds. Instead, the greenhouse, with its stepped rows of planks, provide potted plants for distribution at key points inside and outside the pavilion.

The general area in which the Bagh-i-Eram is situated may be identical with the site called the King's Garden (Bagh-i-Shah) by Ogilby, and by de Bruyn who visited it early in the eighteenth century. It lay to the northwest of the town; as one left the walls through the Iron Gate a very wide avenue stretched for some two thousand paces to the entrance of the King's Garden. The garden itself was 966 paces long and ninety-five paces wide. Its long axis featured an alley twenty paces wide and 620 paces long, which was lined with seventy-two cypress trees. The avenue led to a spacious structure and was renewed and prolonged beyond the building. To the left of the great pavilion was a pool eighty-five paces square. Fountains played at each corner of the pavilion and its grand hall was crowned by a dome. Indeed, the description is quite reminiscent of the proportions and layout of the Bagh-i-Eram.

One more of the large, existing gardens of Shiraz deserves attention. Sometimes known as Afifabad, and sometimes as the Bagh-i-Gulshan, or Flower Garden, the foundation is assigned to 1863 and to the command of Qavam ol-Molk Mirza Ali Muhammad Khan, another of the Qavam-Shirazi family. Inscriptions on the main structure record the names of later members of the family.

The garden is situated a considerable distance to the west of the town. With the elements of imposing pavilion, great pool, and subsidiary pavilion (Plate 70) the type is similar to the Bagh-i-Eram. However, in this case the main structure, erected at great expense from local stone, is quite different (Plate 71). The design blends motifs from the Achaemenid period with tilework of the nineteenth century, and the result is stately but austere and gloomy. With its many rooms and its vast interior reception hall, the structure has lost the spirit of the garden pavilion. While a lower pavilion displays a use of the same materials, it remains light, airy, and attractive even in ruin.

As would be expected, town houses in Shiraz reflect the type that is now so familiar, the open-center pavilion facing south. A characteristic example is notable for the mirror work embellishing all wall surfaces of the open porch (Plate 72). Long vanished and preserved only in a very old photograph, with its name forgotten, its carved and colored bas-reliefs portraying ancient heroes and kings are sheltered in the museum at Shiraz.

Somewhat later in date was the structure known as the Divan Khaneh, or Hall of Audience, belonging to the Qavam-Shirazi family (Color plate IV). Note the characteristic pediment treatment with its scene inspired by antiquity, and the curtain in place to shade the porch on hot days. The plan illustrates another familiar arrangement featuring the pool, the central, decorated path, and the flanking foliage (Plate 73). The main block houses the living quarters, while the kitchen and the servants' quarters are along the street. Of interest is the typical manner of concealing the interior court from view from the street. Again in Shiraz style are the enameled-tile panels showing elegantly attired servants offering fruit and drink to guests (Plate 74).

Just across a narrow lane from the Divan Khaneh is another early nineteenth-century town house belonging to the same Qavam-Shirazi family. This structure, called the Narangistan, is very similar to the Divan Khaneh in plan, except that it is on a much smaller plot of ground and departs from the usual pattern in having the main rooms facing toward the east rather than the south. A series of rooms lines the four sides of the open court. Fine wood paneling and the famed Shiraz inlay work decorate the reception rooms, and the largest of these is completely covered—walls and ceiling—with mirror work. There are pools lined with blue enameled tiles at either end of the court, one provided with a decorative fountain.

Has the roll of the gardens of Shiraz been exhausted? The *Athar-i-'Ajam*, published in Persia in 1896, gives brief accounts of some twenty-eight other Shiraz gardens of local fame. Some of these were named after their owners. Others bore fanciful or descriptive names, for example, the Gardens of the Moat, of Prospect, of Joy, of Saffron, of Pomegranates, and of the Falcon. Typical of this group is a pavilion which stood for more than a century in the heart of crowded Shiraz (Color plate v).

Today the Shirazis display a pride in their home town which is unrivaled in Iran. There is an atmosphere of activity and challenge, with many municipal enterprises completed and more under construction. The townspeople are proud of their fledgling university, of their piped water system and, above all, of the Shiraz Medical Center, multi-million-dollar gift of a native son. Today, as always, the Shirazis are poets one and all. As in the past, they sing the praises of Shiraz in the spring, of its flowers, and of its gardens. Would that these poets might turn to the task of preserving their famed gardens from neglect and decay!

CHAPTER SEVEN

# Gardens North and South

Tabriz was noted for its gardens as early as the visit of Marco Polo in 1300, and when the rulers of the so-called White Sheep dynasty made Tabriz their capital, near the end of the fifteenth century, they adorned it with a marvelous garden called the Hasht Bihesht (Eight Paradises). Some accounts ascribe the garden to the ruler Uzun Hasan and some to his son Yaqub, but all agree on its magnificence. In scale it must have rivaled the much later Hazar Jarib and Farahabad gardens at Isfahan.

Much of the total area was taken up by complexes of palaces and by service quarters. A Venetian merchant who visited Tabriz not long after the garden had been completed, described it in some detail—too much in the way of detail for quotation at this point. He wrote of a series of courts, and of a great pavilion situated at the points of intersection of the cross plan. The pavilion was a single story in height and displayed many rooms around a huge, central hall of audience or reception: built of marble, the walls of the pavilion were decorated with mural paintings, and the central hall was crowned by a great dome, gilded on the exterior. In adjacent pools, ships and boats could be manipulated to demonstrate naval battles.

In later centuries this same site acquired the name of its location, the Bagh-i-Eshratabad, but before the end of the nineteenth century the name had changed again—this time to the Bagh-i-Shimal, or Northern Garden. Curiously enough, this garden was situated on the southern outskirts of the town. During the period of the Qajar dynasty it was customary for the crown prince of the line to act as governor of the province of Azerbaijan, and to be in residence at Tabriz. Several lived in this garden. The plan illustrated, made after an old map, shows the garden reduced in area with only a single large structure remaining (Plate 75).

Some of these Qajar princes showed an interest in embellishing the vicinity of Tabriz. Before the middle of the nineteenth century, one English visitor was charmed with the Garden of Delight, situated a few kilometers from the town. Although lamenting the way its structures were allowed to fall into ruin, he was sufficiently impressed by the natural setting to give his translation of a description of the site made by a local poet: "Behold the sweet groves, beautiful gardens, and flowing streams. Is it not a spot for the abode of heroes? The ground is like velvet and the air breathes perfume—you would say that the rose had imparted its scent to the water of the rivulets—the stalk of the lily bends under the weight of the flower, and the whole grove is charmed with the fragrance of the rose bush—from this moment time has no meaning. May the flower beds of these banks resemble the bowers of Paradise."

At the present day there are two charming gardens in the vicinity of Tabriz; they are situated some distance to the east of the town and just to the south of the main highway from Tabriz to Tehran. The grander is called the Shah Gol, Shah Goli, or Shah Kol—all versions of the same term, a combination of Persian and Turkish words meaning the Royal Pond (Plate 77). The principal feature is the artificial lake, just over seven hundred feet on a side. According to a local history of Tabriz, it was built by an unnamed king in 1785—probably it was made at a considerably earlier date. This same source goes on to say that the lake was cleaned out, the terraces built, and the pavilion erected prior to the middle of the nineteenth century.

The similarity to the Bagh-i-Takht at Shiraz is immediately apparent, although here the scale is much greater. The vast pool was not excavated from a level spot of ground, but instead the northern side was built up by moving great masses of earth into place. Foliage forms a screen which conceals all the earth along the edge: as a result, as one looks across from the levels of the terraces, the spreading sheet of water appears to hang suspended far above the valley beyond.

As in the Bagh-i-Takht at Shiraz, the water supply comes from a spring which gushes from the living rock above the highest terrace. From reservoirs, the water is conducted into five channels which create as many waterfalls at each change of level. Each terrace has a row of poplars along its back wall, and willows along the water channels. The rest of the area is divided into numerous plots covered with fruit trees, and, according to the seasons, crops of clover or alfalfa grow under the shade of the trees.

The causeway leads out to a pavilion of unusually elaborate plan. The central hall is a generous octagon, some twelve meters across, entered on each side of the structure. However, each corner angle is broken up into a series of small chambers, niches, and reveals. Originally the pavilion must have been crowned with a dome. Although the dome has vanished, the structure itself was put into good repair a few years ago when the garden was rented to serve as an outdoor café for the residents of Tabriz. No reports on the success of this venture are available. However, such a site deserves an appreciative clientele, and it would be most unfortunate if this fine example of hillside and lake garden should vanish because of indifference and neglect.

Just a short distance from the Shah Gol is the garden of Fathabad. The heart of the garden is almost completely hidden within acres of orchard, but the long water axis ties the whole together in Persian fashion and all is in harmony, except for the modern house which has replaced an earlier pavilion (Plate 76). Starting from the upper, and southern, end of the garden, the central avenue is marked by changes in level and by variety in the shape of pools. In one section a series of grass plots borders the water channel, and these plots are edged in turn by rows of potted geraniums. The plan reaches its climax at the huge stone-lined pool—very deep and bordered around by massive, ancient trees. The northern side of this pool is built up above ground level—a feature noted at the Shah Gol. No aspects of this garden are particularly elegant or striking, and yet the total impression is one of great calm and charm.

These two examples must suffice for the gardens of northwestern Iran. Since that part of the country enjoys a considerably heavier annual rainfall than the rest of the plateau, flowers, trees, and fields will flourish without irrigation, and there may have been less compulsion to exploit water and shade than in more arid sections.

If the traveler leaves Tabriz and follows the road to Tehran, he can keep going for a long time and a great distance. From Tehran a highway goes to Qum, and at this point a road branches southeast toward Kashan, noted for its roses. Just upon coming in sight of historic Kashan, the traveler looking to the right of the road would see the towering cypresses of the garden at Fin. A visit to this spot would be the high point of any trip, for Fin is by far the best example of the large, formal Persian garden.

In 1504 Fin, a suburb of Kashan, was the site of a reception in honor of the first Safavid ruler, Shah Ismail. After 1587 Shah Abbas erected structures at the site, and we know that his namesake, Shah Abbas II, paid a visit to the garden in 1659. However, this earlier work seems to have vanished, and all the existing structures are ascribed to Fath Ali Shah, who ruled from 1797 until 1834.

Persians associate the site with an historical incident, rather than its own beauty, for in 1852 royal executioners put to death the honest, capable prime minister of this period, Amir-i-Kabir. The warrant was carried out by opening the victim's veins in one of the rooms of the bath—the complex of rooms in the lower left corner of the enclosure (Plate 78). For some seventy-five years after this mournful incident the garden was allowed to deteriorate, with the buildings slowly collapsing from neglect. Then, in 1935, the Bagh-i-Fin was named a national monument of Iran, and vitally needed repairs were undertaken. Much more in the nature of reconstruction could be done, and one day the stately central pavilion may appear in its early elegance.

Fin merits close attention because it is an admirable example of the monumental royal garden, and because it is the very epitome of the Persian garden—this single example displaying all the most desired features and elements. At the risk of restating remarks made in earlier chapters, the meaning of Fin to the Persians should be stressed. The garden expresses a series of accentuated contrasts between the arid, inhospitable landscape outside the walls and the lush foliage within. Outside, water is scarce and precious; here it flows with superabundance (Plate 79) to produce a dense jungle of growth. The monotone of the landscape is replaced by the colors of foliage, of flowers, of blue tiles, of fountains, and of painted plaster and woodwork. Axial symmetry contrasts with areas of almost impenetrable growth. The plan of Fin calls sharply to mind the Persian garden carpet, for all the elements of multiple channels, orchards, flowers, and pavilions are present in a similar relationship.

At Fin all the channels are lined, sides and bottom, with blue faïence tiles so that the very water seems bright and gay until it flows into one of the larger pools, lined with great trees (Plate 80). The largest pool mirrors the remains of the central pavilion, ascribed to Fath Ali Shah. A sketch made a hundred years ago (Plate 81) shows its original aspect, while the artist records that the rooms were decorated with murals, including one showing the ruler surrounded by twenty of his sons. This painting was yet another version of ones at Tehran, and it seems quite reasonable that the father of so many children should want to be sure that adequate records were left of his achievement.

The existing complex contains too many sections for description in any short account. Several entrances pierce the very high enclosing walls. The principal one—on the north—was approached by a ceremonial avenue bordered by tall trees, and on the west a smaller portal allowed the royal suite to ride directly into the stables. The garden palace had its separate anderun, or women's quarters, and the complex of rooms adjacent to the stable area may have served this function. The bath, already mentioned, contained separate rooms and pools for hot and cold water. Along the walks leading through the garden, fountains spout from lead nozzles, and the pool to the south of the central pavilion displays 160 such nozzles. Stately lines of cypresses and plane trees border these same walks, and help to create the perfect atmosphere of the Persian garden (Plate 82).

From Kashan the highway goes on to Yazd and then to Kirman. About forty kilometers beyond Kirman is the shrine of Mahan, pronounced locally Mahun. Its structures are grouped around the tomb of Sayyid Shah Ni'matullah, a poet, mystic, and saint, and the founder of a religious brotherhood. Active in the fifteenth century, he lived in honor and respect at Samarqand, Herat, and Yazd, and then passed the last twenty-five years of his life at Mahan. Thousands of Persians belong to his Ni'matullah order, and many pious Muslims make the *ziaret*, or visitation, at his tomb. The history of this poet-saint and his continuing influence while interesting, is less relevant to a study of gardens than is the environment of the shrine. This shrine may be taken as typical of many throughout Iran, perhaps of many hundred, where the combination of a pool and a few trees serves to create an atmosphere of tranquillity and of isolation from the frustrations of the active world.

At Mahan the shrine includes a succession of three open courts, with surrounding structures, on a central axis. Each court has its pool and trees (Plate 83). The forecourt, with its minaret-crowned portal, is given the most elaborate treatment, centering upon a cross-shaped pool (Plate 84). Around the pool are two sorts of evergreens: stately old cypresses, and pine trees with all their lower branches trimmed away. Rose bushes inject a gayer note into the dark tones of trees and shadowed water. The pool itself provides a most unusual variation of the standard type. A smaller octagonal pool at the axis of the cross-shaped pool, and at higher level, responds in a different manner to stray breezes, while the composition of water builds up into a conical form when its fountain plays.

Mahan, Kirman, and Yazd lie along the fringes of the vast deserts of Iran. A single motorable road, leading northwest from Yazd, pierces these desolate wastes in the direction of Torbat-i-Hayadari and Mashhad. About midway through the desert the hardy traveler comes upon Tabas, a true oasis town. Rainfall is very rare here and no streams or rivers are to be found, but artesian springs bubble forth to create life in the midst of a barren waste. The presence of ancient mosques, minarets, and a caravanserai bear witness to the great age of the town; and the existence of ruined Qajar palaces indicates that the long trip to this site was considered sufficiently rewarding. Of primary interest is the manner in which the springs of water determine the plan of the town. The site of the spring is, of course, a garden, formerly owned by the crown and now by the municipality. The natural slope from this site establishes the long axis of the town, and immediately below the garden is the main square. From this square, water channels mark the main axis, and also lead off at other angles to flanking sectors of the town.

At the garden all attention is concentrated upon the springs, or rather upon a level just below the actual springs where the abundant water supply rushes out of three jets into a pool of considerable size (Plate 85). Walks border the pool, but the stress is not upon formal arrangement but upon maximum contrast between water and foliage. Surrounding the pool are the dense masses of cypresses, willows, and plane trees, relieved by a few towering palm trees. Below the pool the treatment is more open, featuring rose beds and a seasonal display of white petunias. Nightingales trill in the dense shade, while just a score of meters away the activity of the town centers in the sun-drenched square: again the maximum contrast between the worlds of action and of contemplation has been achieved.

# Bibliography

Allemagne, Henry René d'. *Du Khorassan au pays des Backhtiaris.* Paris: Hachette, 1911.

Babar, Emperor of Hindustan. *Memoirs of Zehir-Ed-Din Muhammed Baber, Emperor of Hindustan.* Translated by John Leyden and William Erskine. London: Longman, Rees, Orme, Brown, and Green, 1826.

Beaudouin, Eugène-Elie. "Isfahan sous les grands shahs (XVIIe siècle)." *Urbanisme,* no. 10 (1933).

Binder, Henry. *Au Kurdistan, en Mésopotamie et en Perse.* Paris: Quantin, 1887.

Binning, Robert B. M. *A Journal of Two Years' Travel in Persia.* London: W. H. Allen, 1857.

Blunt, Wilfrid. "The Persian Garden under Islam." *Apollo* 103 (1976): 302–6.

Browne, Edward Granville. *A Literary History of Persia.* Vol. 3. Cambridge: The University Press, 1928.

Browne, Edward Granville. *A Year amongst the Persians: Impressions . . . Received during . . . 1887–1888.* New ed. Cambridge: The University Press, 1926.

Bruyn, Cornelis de. *Travels into Muscovy, Persia, and Part of the East Indies.* London: A. Bettesworth, 1737.

Bruyn, Cornelis de. *Voyages de Corneille Le Brun par la Moscovie, en Perse, et aux Indes Orientales.* Amsterdam: Frères Wetstein, 1718.

Buckingham, James Silk. *Travels in Assyria, Media, and Persia.* London: H. Colburn, 1829.

Chardin, Sir John. *Voyages de monsieur le chevalier Chardin, en Perse, et autres lieux de l'Orient.* Amsterdam: Chez Jean Louis de Lorme, 1711.

Coste, Pascal. *Monuments modernes de la Perse, mesurés, dessinés, et décrits.* Paris: A. Morel, 1867.

Crowe, Sylvia; Haywood, Sheila; Jellicoe, Susan; and Patterson, Gordon. *The Gardens of Mughul India: A History and Guide.* London: Thames and Hudson, 1972.

Curzon, George N. *Persia and the Persian Question.* London: Longmans, Green, 1892.

Dieulafoy, Jane. *La Perse, la Chaldée et la Susiane*. Paris: Hachette, 1887.

Eastwick, Edward B. *Journal of a Diplomate's Three Years' Residence in Persia*. London: Smith, Elder, 1864.

Feuvrier, Joannes. *Trois ans à la cour de Perse [par] Docteur Feuvrier*. New ed. Paris: Maloine, 1906.

Flandin, Eugène, and Coste, Pascal. *Voyage en Perse: Perse moderne*. Paris: Gide et J. Baudry, 1854.

Fowler, George. *Three Years in Persia; with Travelling Adventures in Koordistan*. London: H. Colburn, 1841.

Fraser, James B. *Travels and Adventures in the Persian Provinces on the Southern Banks of the Caspian Sea*. London: Longman, Rees, Orme, Brown, and Green, 1826.

Fryer, John. *A New Account of East-India and Persia*. London: R. Chiswell, 1698.

González de Clavijo, Ruy. *Clavijo, Embassy to Tamerlane, 1403–1406*. Translated by Guy Le Strange. London: G. Routledge & Sons, 1928.

Haentzsche, J. C. "Paläste Schah Abbas I. von Persien in Masanderan." *Zeitschrift der Deutschen Morgenländischen Gesellschaft* 18 (1864): 669–79.

Hanway, Jonas. *An Historical Account of the British Trade over the Caspian Sea; with the Author's Journal of Travels from England through Russia into Persia*. 2d ed. London: T. Osborne, 1754.

Hasan ibn Hasan, called Fasa'i, and Tabib, Shirazi. *Fars-nameh i nasiri* [A History and Geography of Fars]. Tehran: A.H. 1313 (A.D. 1895).

Herbert, Sir Thomas. *Some Yeares Travels into Divers Parts of Asia and Afrique*. 2d ed. London: I. Blome and R. Bishop, 1638.

Hommaire de Hell, Xavier. *Voyage en Turquie et en Perse*. Paris: P. Bertrand, 1854–60.

Ibn Arabshah, Ahmad ibn Muhammad. *Tamerlane; or, Timur, the Great Amir*. Translated by J. H. Sanders. London: Luzac, 1936.

Jackson, Abraham V. W. *Persia Past and Present*. New York: Macmillan, 1906.

Jaubert, Pierre Amédée. *Voyage en Arménie et en Perse, fait dans les années 1805 et 1806*. Paris: Pélicier, 1821.

Kaempfer, Engelbert. *Amoenitatum exoticarum politico-physico-medicarum, fasciculi V, quibus continentur variae relationes, observationes et descriptiones rerum Persicarum et ulterioris Asiae*. Lemgo: H. W. Meyer, 1712.

Keppel, George Thomas. *Personal Narrative of a Journey from India to England . . . in the Year 1824*. London: H. Colburn, 1827.

"Landscape Architecture and Gardening of the Mughals." *Marg: A Magazine of the Arts*, Dec. 1972, 3–8, 12–48.

Lycklama à Nijeholt, Tinco Martinus. *Voyage en Russie, au Caucase et en Perse . . . pendant les années 1866, 1867 et 1868.* Paris: A. Bertrand, 1872–75.

MacDougall, Elisabeth B., and Ettinghausen, Richard, eds. *The Islamic Garden.* Dumbarton Oaks Colloquium on the History of Landscape Architecture, vol. 4. Washington: Dumbarton Oaks, Trustees for Harvard University, 1976.

Melgunov, Gregorii. *Das südliche Ufer des Kaspischen Meeres; oder, die Nordprovinzen Persiens.* Leipzig: L. Voss, 1868.

Morgan, Jacques Jean Marie de. *Mission scientifique en Perse.* Vol. 4, *Recherches archéologiques.* Paris: E. Leroux, 1896–97.

Morier, James. *A Journey through Persia, Armenia, and Asia Minor, to Constantinople, in the Years 1808 and 1809.* London: Longman, Hurst, Rees, Orme, and Brown, 1812.

Muhammad Nasir, called Fursat, and Mirza Aqa, Husaini. *Asar i Ajam* [an illustrated encyclopaedic work on the southeastern district of Persia]. Bombay: A.H. 1314 (A.D. 1896).

Mumford, John Kimberly. "Glimpses of Modern Persia." *House and Garden* 2 (1902): 175–91, 360–73, 429–36.

Ogilby, John. *Asia, the First Part: Being an Accurate Description of Persia, and the Several Provinces thereof.* London: The Author, 1673.

Olearius, Adam. *Relation du voyage d'Adam Olearius en Moscovie, Tartarie, et Perse.* New ed. Paris: I. du Puis, 1666.

Ouseley, Sir William. *Travels in Various Countries of the East: More particularly Persia.* London: Rodwell and Martin, 1819–23.

Porter, Sir Robert Ker. *Travels in Georgia, Persia, Armenia, Ancient Babylonia . . . during the Years 1817, 1818, 1819, and 1820.* London: Longman, Hurst, Rees, Orme, and Brown, 1821–22.

Rabino, Hyacinth Louis. *Mazandaran and Astarabad.* London: Luzac, 1928.

Sackville-West, V. "Persian Gardens." In *The Legacy of Persia*, edited by A. J. Arberry. Oxford: Clarendon Press, 1953.

Sarre, Friedrich. *Denkmäler Persischer Baukunst.* Berlin: E. Wasmuth, 1901–10.

Sarre, Friedrich. "Reise in Mazandaran (Persien)." *Zeitschrift der Gesellschaft für Erdkunde zu Berlin* (1902) no. 2: 99–111.

Sercey, Édouard, comte de. *Une ambassade extraordinaire: La Perse en 1839–40.* Paris: L'Artisan du livre, 1928.

Shahzada Nader Mirza. *Tarikh va joghrafi dar al-saltana Tabriz* [History and Geography of the Capital, Tabriz]. Tabriz: A.H. 1323 (A.D. 1905).

Sharaf ad-Din Ali Yazdi. *The History of Timur-Bec . . . Written in Persian by Cherefeddin Ali.* Translated into French by Petis de la Croix, into English by J. Darby. London: J. Darby, 1723.

Stuart, Charles. *Journal of a Residence in Northern Persia and the Adjacent Provinces of Turkey by Lieut.-Colonel Stuart.* London: R. Bentley, 1854.

Tavernier, Jean Baptiste. *The Six Voyages of John Baptista Tavernier . . . through Turky into Persia, and the East Indies.* London: R.L. and M.P., 1678.

Texier, Charles. *Description de l'Arménie, la Perse, et la Mésopotamie.* Paris: Firmin Didot frères, 1842–52.

Thévenot, Jean de. *The Travels of Monsieur de Thévenot into the Levant.* London: H. Faithorne, J. Adamson, C. Skegnes, and T. Newborough, 1687.

"Travels of a Merchant in Persia." In *A Narrative of Italian Travels in Persia in the 15th and 16th Centuries*, translated and edited by Charles Grey, pp. 139–207. Hakluyt Society: Works, 1st series, vol. 49. London: The Hakluyt Society, 1873.

Valla, Pietro della. *Voyages de Pietro della Valle . . . dans la Turquie, l'Égypte, la Palestine, la Perse, les Indes Orientales et autres lieux.* New ed. Rouen: R. Machuel, 1745.

Whitaker, Charles H. *Bertram Grosvenor Goodhue—Architect and Master of Many Arts.* New York: American Institute of Architects, 1925.

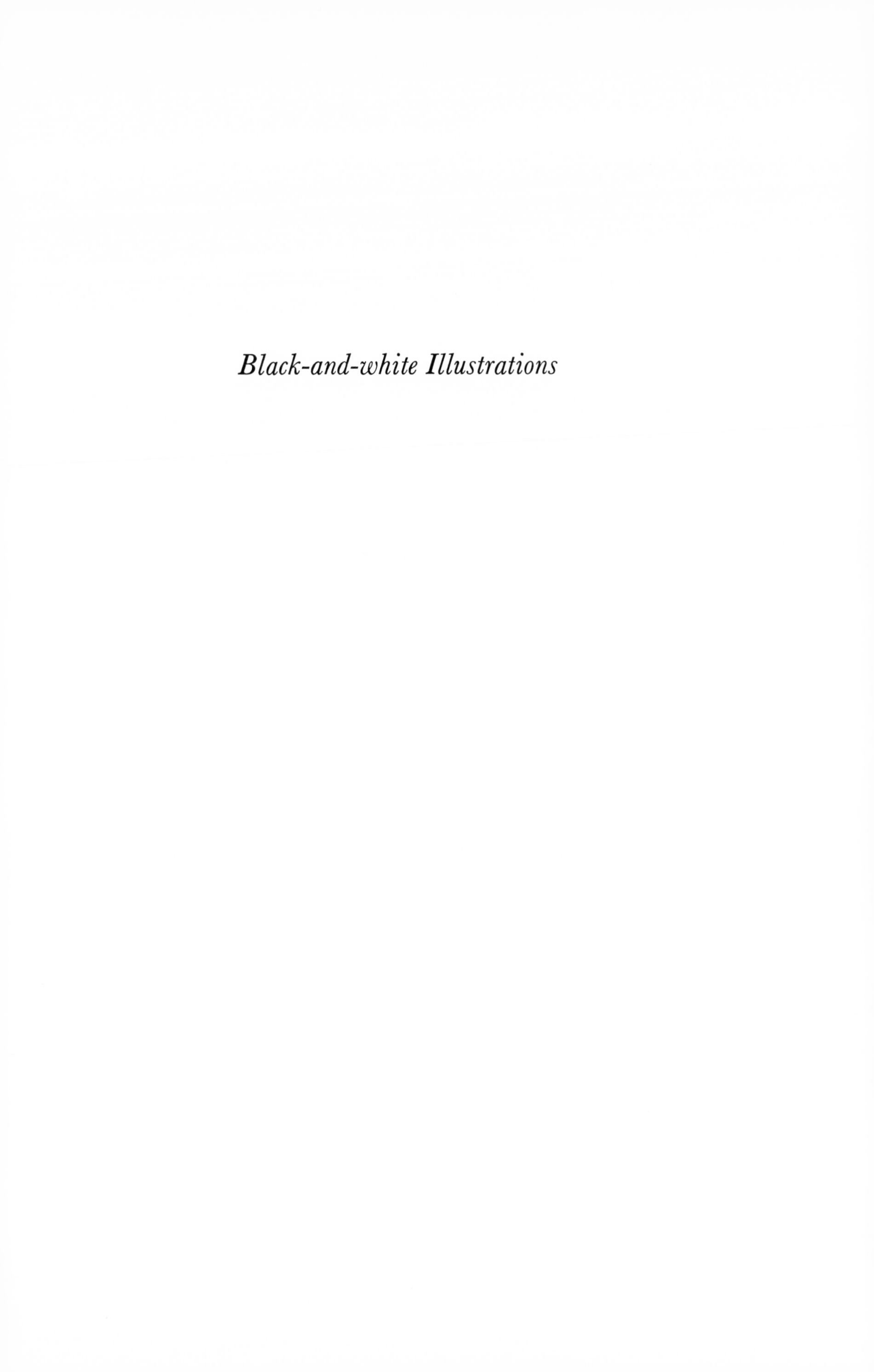

*Black-and-white Illustrations*

1. A qanat-supplied great pool in a royal garden at Tehran

2. Book cover, with flowers painted in an embroidery style
by Muhammad Ibrahim Saifullah al-Husayni, 1836 (photo: ASI)

3. Bokhara embroidery, 19th century, Metropolitan Museum of Art (photo: same)

4. Flower painting; pink roses in the style of Shiraz

5. Flower painting; full-blown rose

6. Large garden carpet, 18th century (photo: courtesy of Arthur Upham Pope)

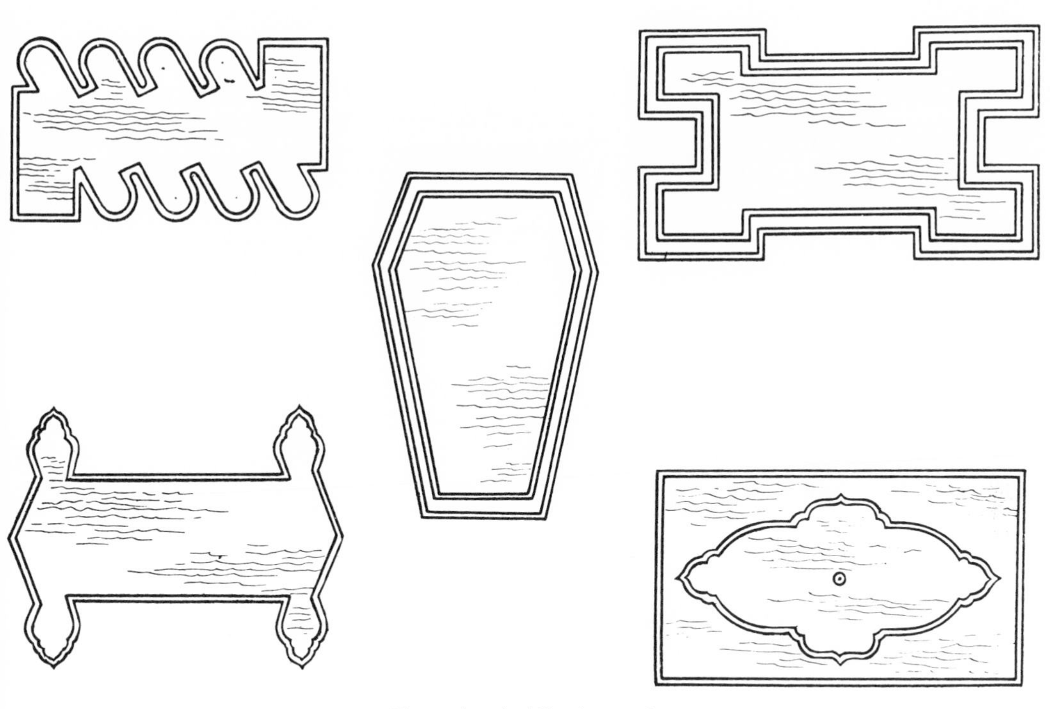

7. Plans of typical Persian pools

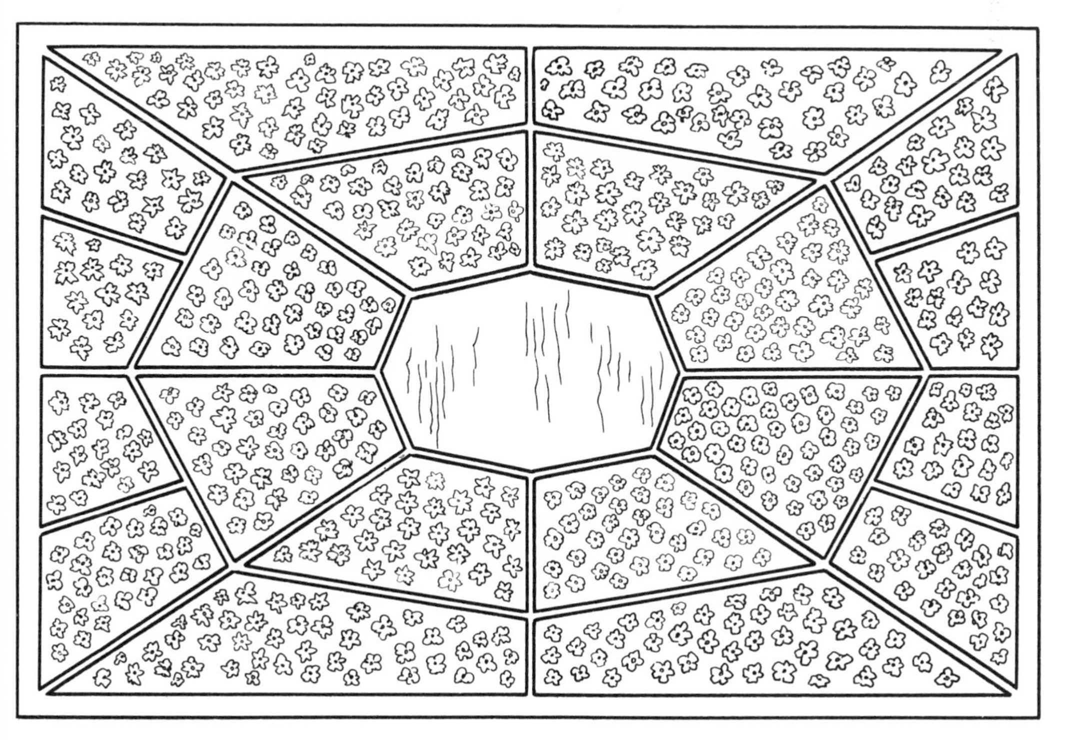

8. Example of pool with imitated carpet pattern using sectioned, flower-strewn areas

9. *Prince in Autumnal Garden*, miniature, early 16th century, Freer Gallery of Art, ms. 32.50, f. 40v (photo: same)

10. *The Garden of the Fairies*, miniature, early 16th century, Freer Gallery of Art, ms. 50.2 (photo: same)

11. *Alexander*, from an episode in the *Khamseh* of Nizami, miniature, Shiraz, mid-16th century, Freer Gallery of Art, ms. 08.280 (photo: same)

12. *Young Prince in Garden Kiosk*, miniature, late 16th century, Metropolitan Museum of Art, ms. 11.39.1 (photo: same)

13. Episode from the *Haft Awrang* of Jami, miniature, mid-16th century, Freer Gallery of Art, ms. 46.12-52 (photo: same)

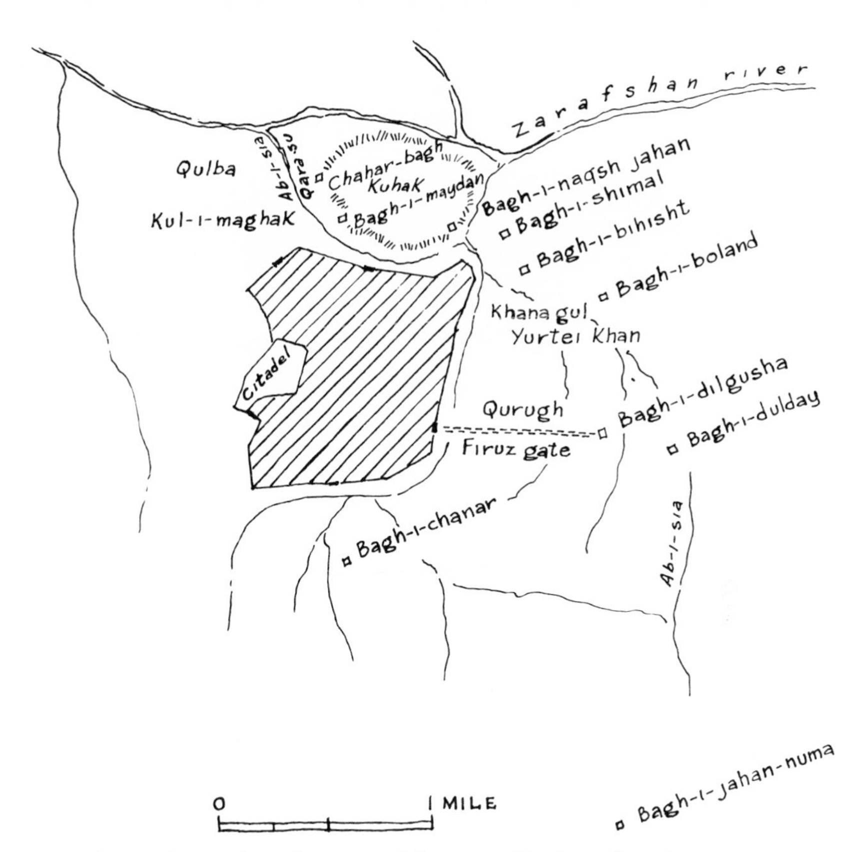


14. Map showing gardens and meadows around Samarqand in the 15th century

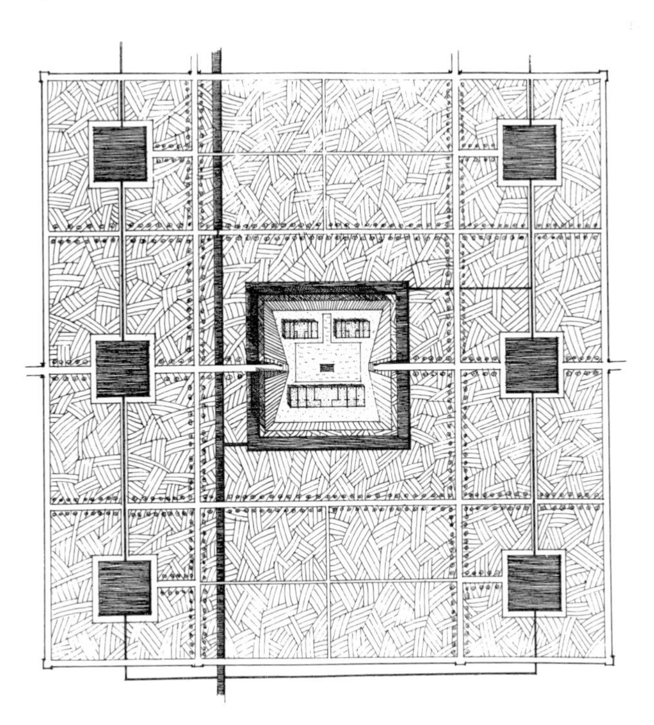

15. Reconstructed plan of a garden at Samarqand, probably that known as the Rose Garden

16 & 17. *Timur within a Royal Garden,* frontispiece of the *Zafar nama* of Sharaf ad-din Ali Yazdi, miniature, Walters Art Gallery, ms. T.L.6.1950, ff. 82v, 83 (photo: same)

18. *Babur Giving Instruction for the Layout of the Bagh-i-Vafa*, miniature, Victoria and Albert Museum, ms. I.M.1913.276A, 276 (photo: same)

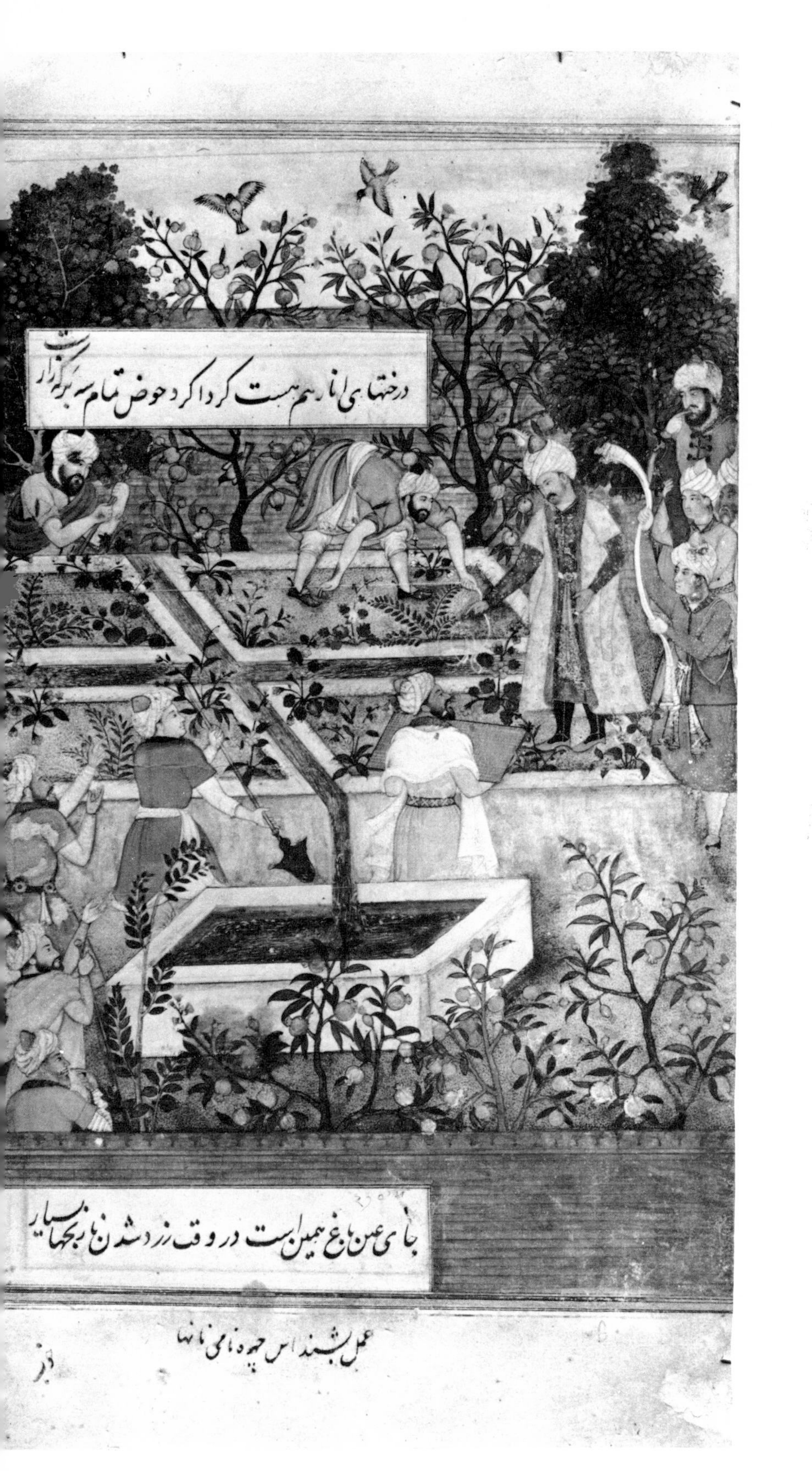
درختهای انار هم هست گرداگرد حوض تمام سه برگه زار است
جای عجب باغ همین است در وقت زرد شدن برگها بسیار
عمل بشنداس چهره نامی نانها

19. *The Bagh-i-Vafa*, miniature, British Library, ms. Or 3714, f. 173v (photo: same)

20. *Babur Inspecting the Bagh-i-Vafa*, miniature, British Library, ms. Or 3714, f. 180v (photo: same)

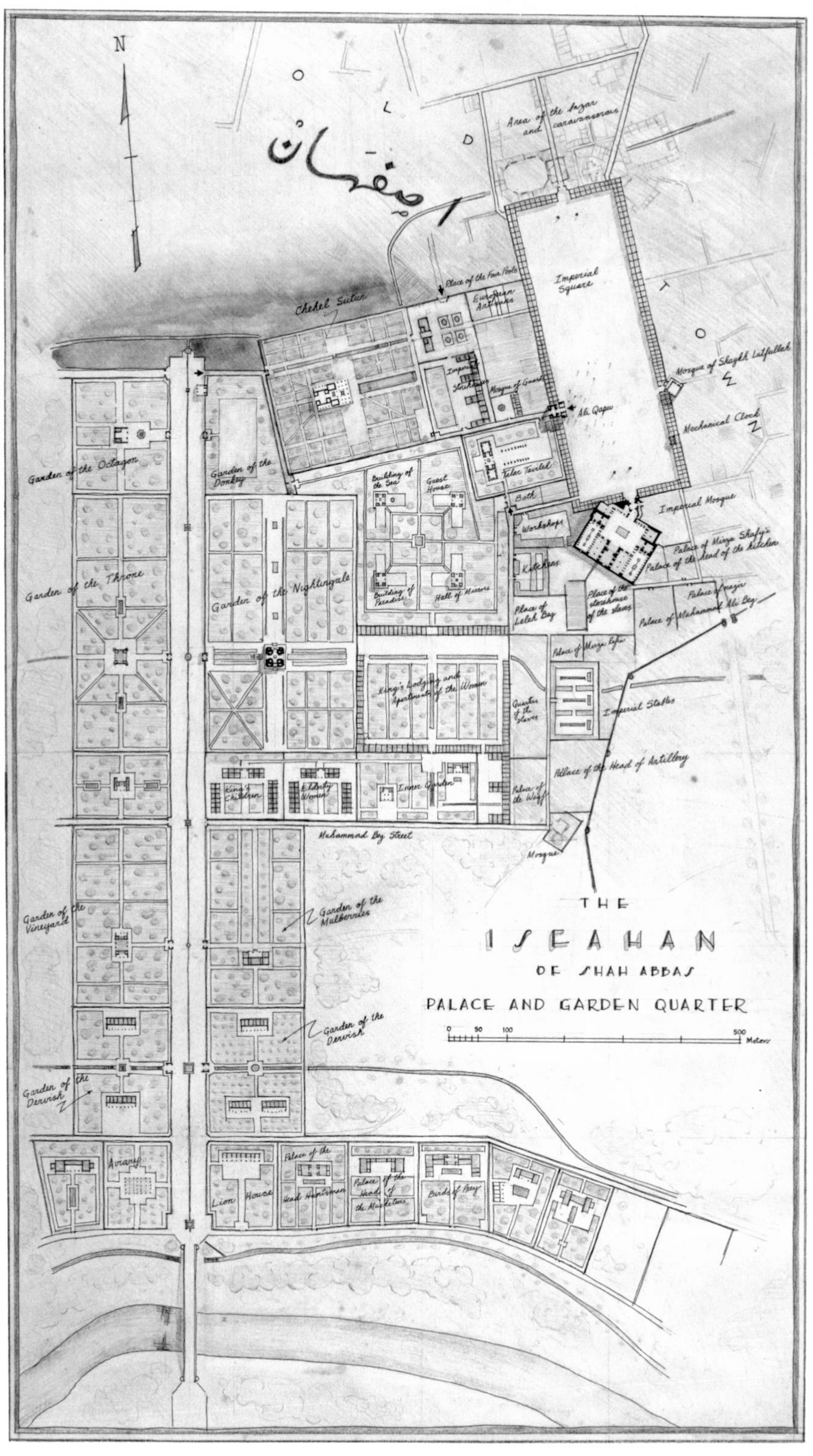


21. Reconstructed plan of the palace and garden quarter of Isfahan in the Safavid period

22. The Maidan-i-Shah at Isfahan as seen from the Ali Qapu; the Masjid-i-Shaykh Luftullah is in the background (photo: Library of Congress)

23. Dome of the Masjid-i-Shah at Isfahan (photo: Library of Congress)

24. Dome of the Masjid-i-Shaykh Lutfullah (photo: Library of Congress)

25. The Ali Qapu as seen from the roof of the Masjid-i-Shah (photo: Library of Congress)

26. *Evening Reception at the Talar-i-Tavileh, Adjacent to the Ali Qapu*, engraving, Kaempfer

27. *Interior of the Now-vanished Sar Pushideh in the Royal Garden Area at Isfahan*, engraving, Coste, pl. XLIV

28. *View from the Porch of the Chehel Sutun; Dome and Minarets of the Masjid-i-Shah in the Distance*, lithograph, Flandin, pl. LI

29. *One of the Now-vanished Palaces once Situated along the Chahar Bagh*, lithograph, Hommaire, pl. XCI

30. *Royal Reception at the Garden Pavilion at Asadabad, Adjacent to the Chahar Bagh*, engraving, Kaempfer

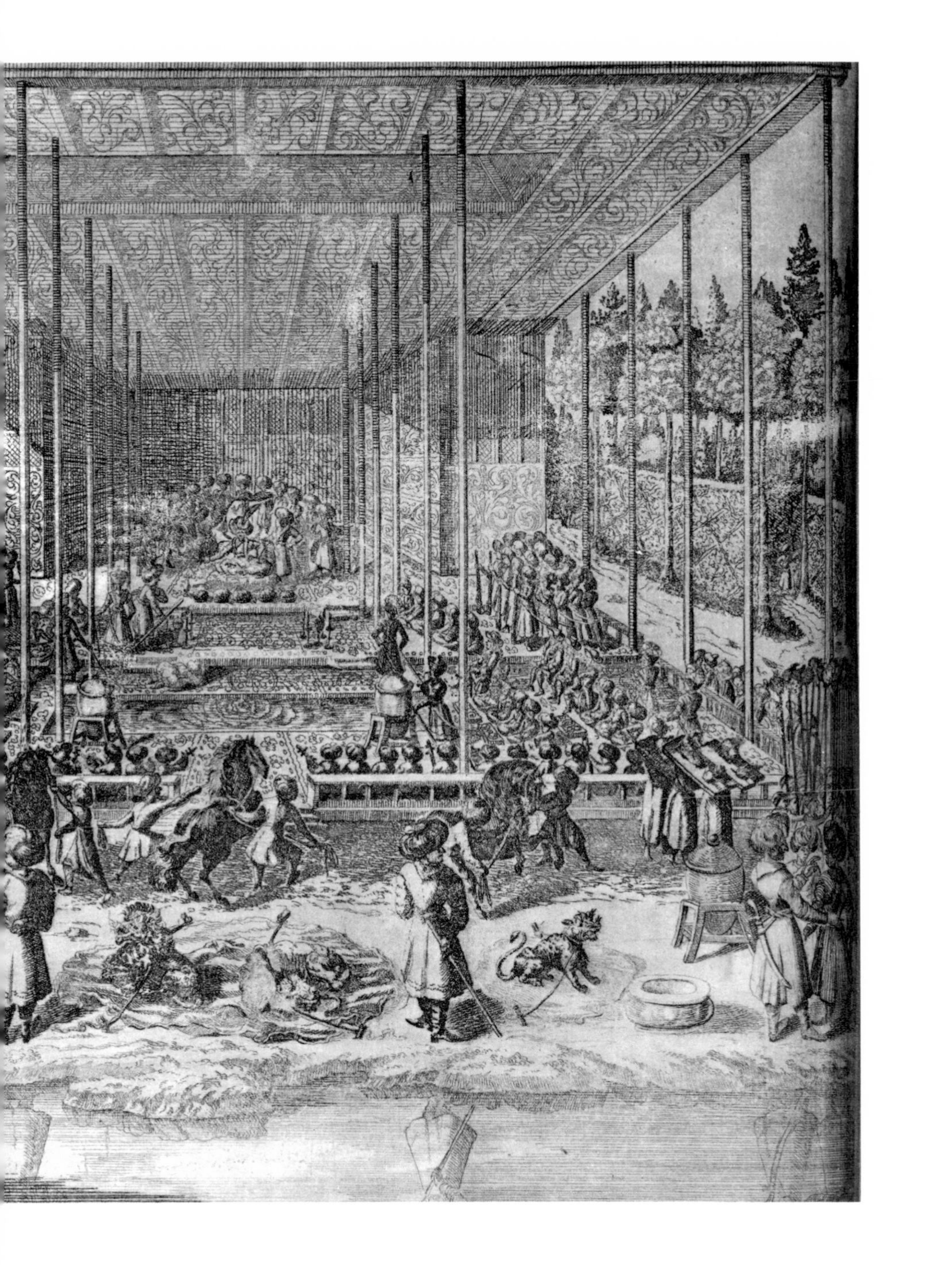

31. *Exterior of the Hasht Bihesht*, engraving, Coste, pl. XXXVI

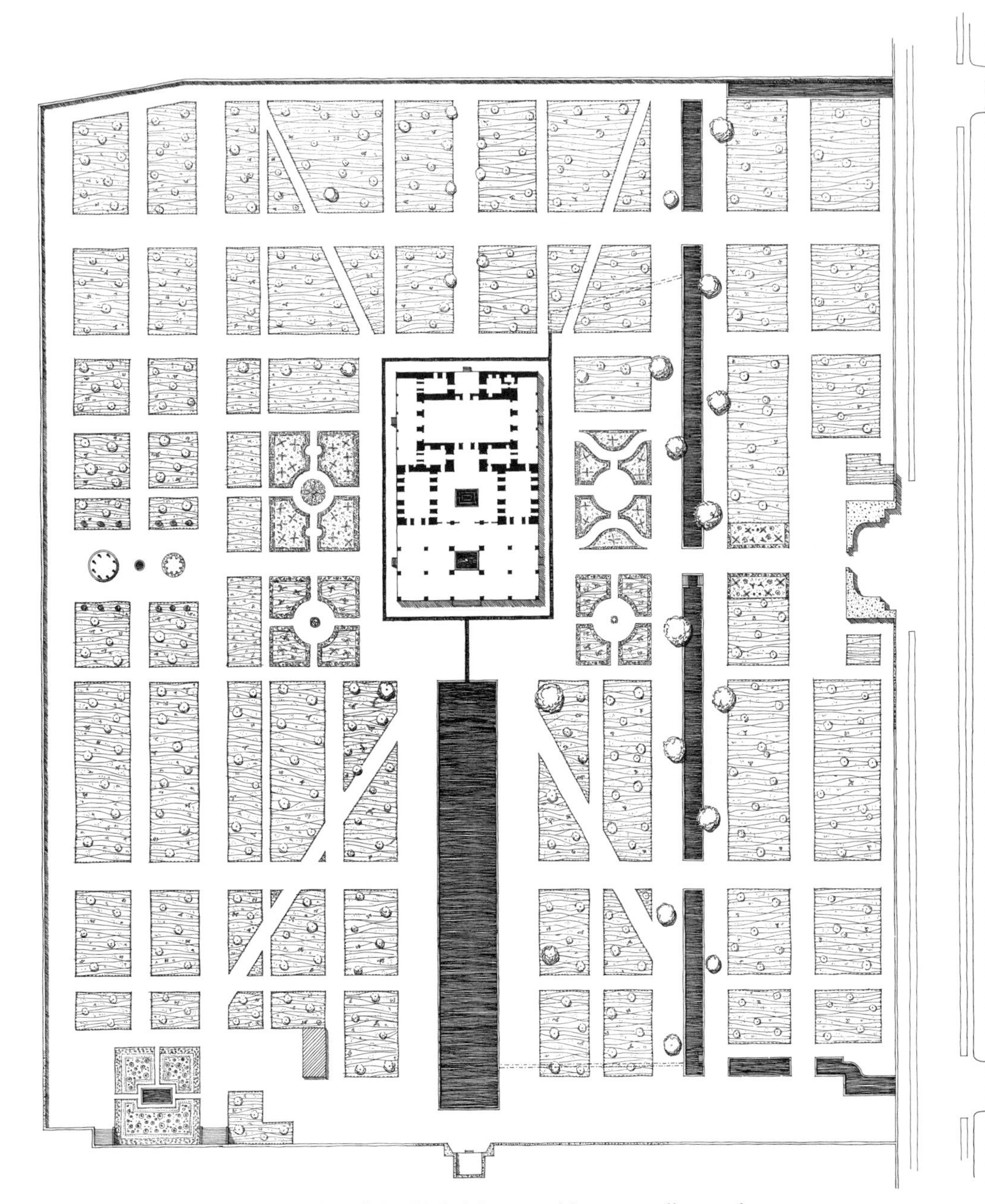

32. Plan of the Chehel Sutun and its surrounding garden

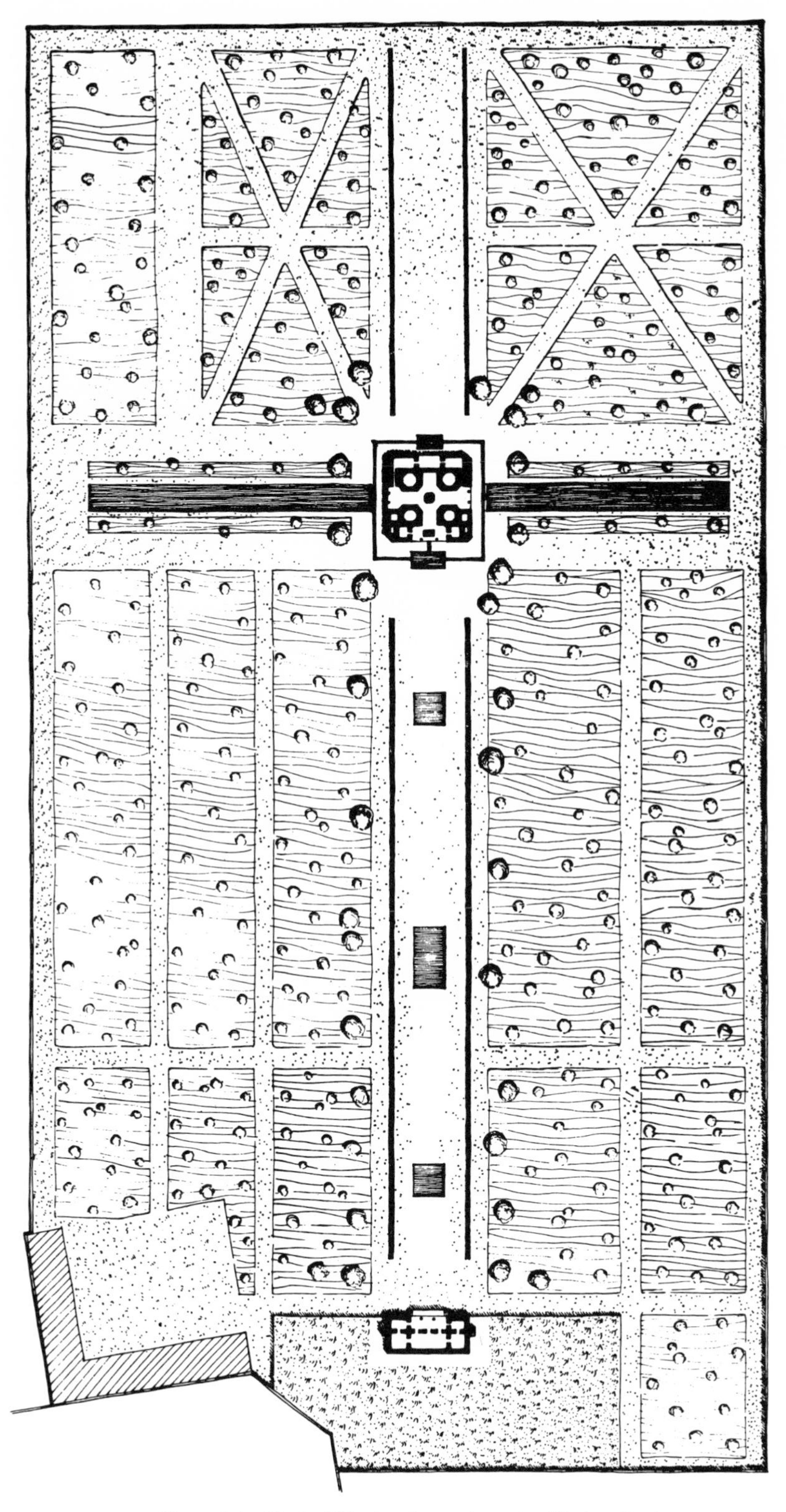

33. Plan of the Hasht Bihesht, situated on the Chahar Bagh and adjacent to the Chehel Sutun

34. *Central Hall of the Hasht Bihesht*, engraving, Coste, pl. XXXVIII

35. Façade of the Madrasa Mader-i-Shah, facing the inner courtyard (photo: Library of Congress)

36. Dome of the Madrasa Mader-i-Shah (photo: Library of Congress)

37. *Perspective of the Now-vanished Ayina Khaneh; the Siosehpol Bridge Appears in the Background,* engraving, Coste, pl. XXXIII

38. *Garden Palace of Shah Abbas at Sari on the Caspian Coast*, lithograph, Hommaire, pl. LXXVIII

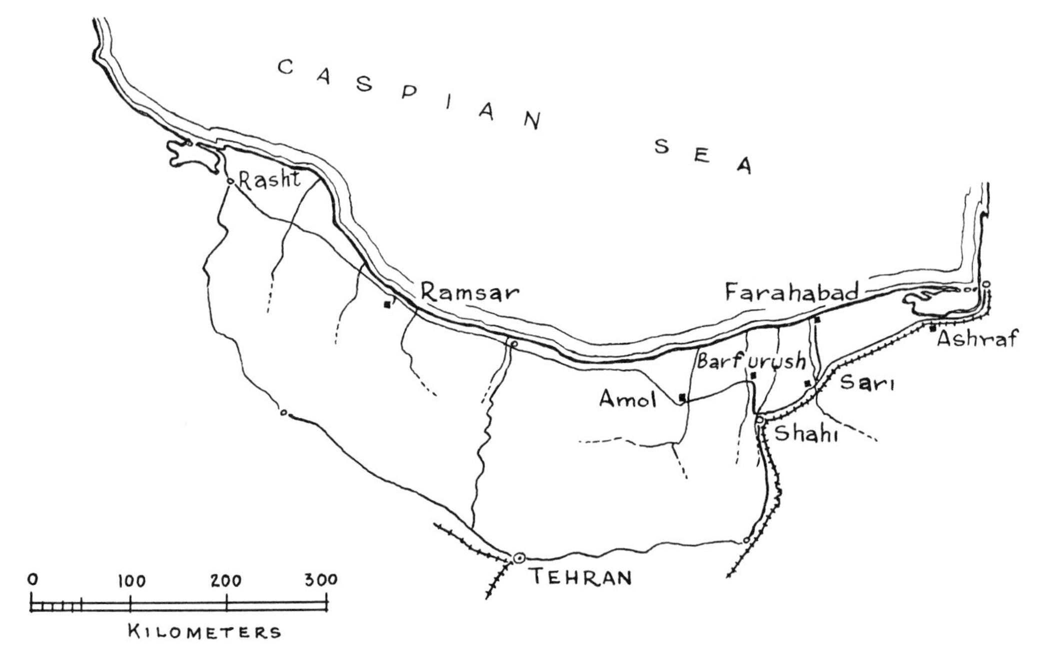


39. Sketch map of sites of royal gardens along the Caspian Sea

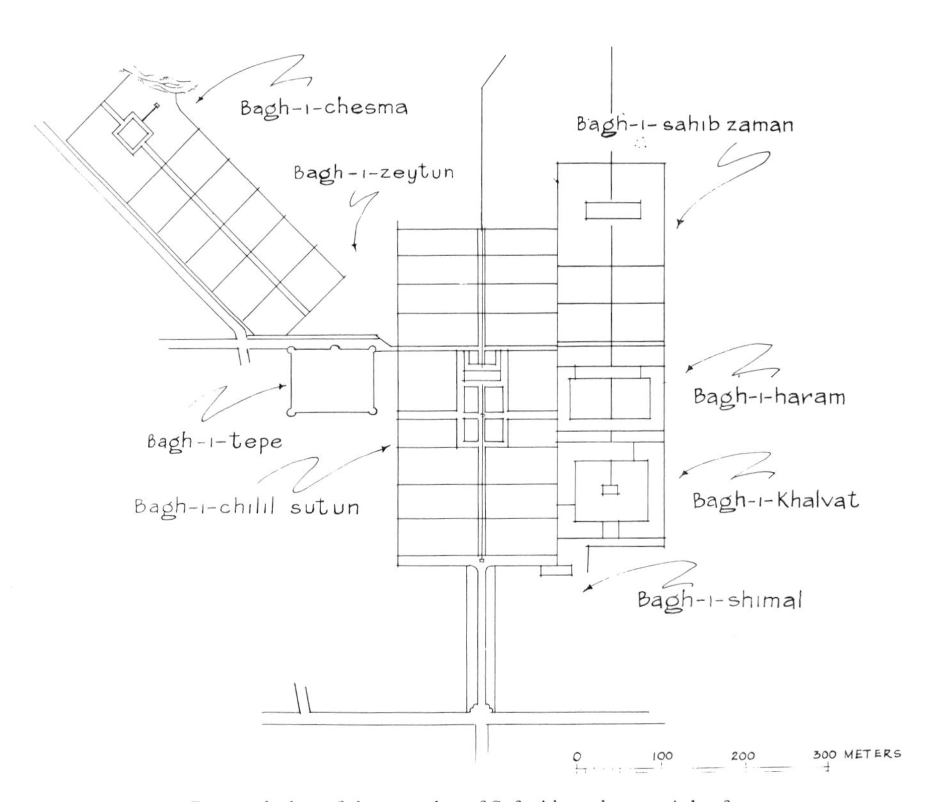


40. Restored plan of the complex of Safavid gardens at Ashraf, known collectively as the Bagh-i-Shah

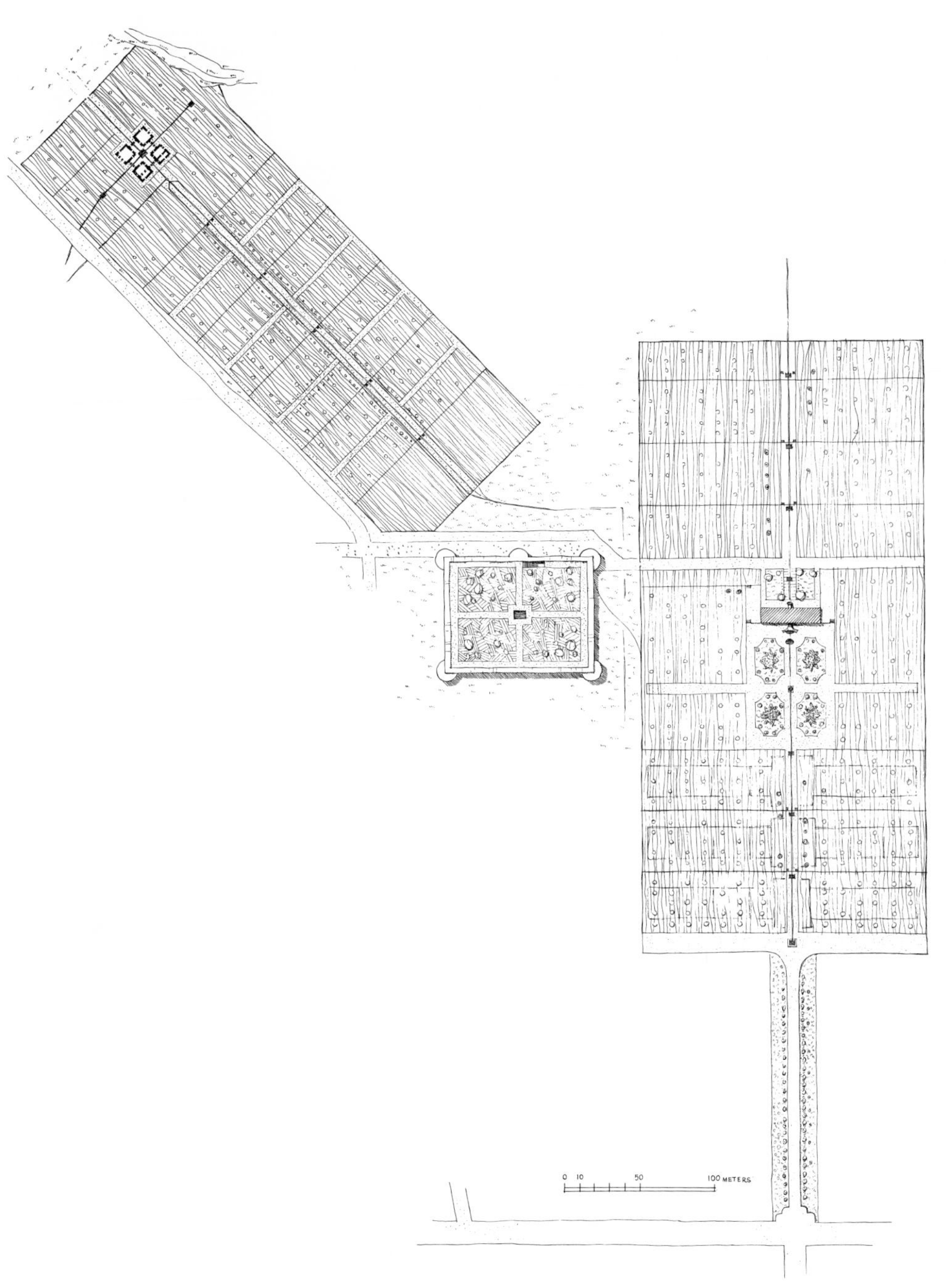


41. Plan of the surviving gardens at Ashraf

42. *Now-vanished Garden Palace in the Chehel Sutun Area at Ashraf*, lithograph, Hommaire, pl. LXXX

43. Safiabad, a promontory crowned by a garden palace of the periods of Shah Abbas and Shah Safi I

44. *The Gate of Happiness to the Gulistan Palace at Tehran*, painting, ca. 1860 (photo: ASI)

45. *Inner Courtyard of the Gulistan Palace*, painting (photo: ASI)

46. *Inner Courtyard of the Rebuilt Gulistan Palace*, painting (photo: ASI)

47. *Reception Hall, with Marble Throne, of the Gulistan Palace*, lithograph, Flandin, pl. XXXI

48. *Typical Tehran House of the 19th Century with Courtyard, Pool, and Garden*, lithograph, Hommaire, pl. LXI

49. *The Building of the Sun within the Complex of the Gulistan Palace*, painting (photo: ASI)

50. View of the royal palace at Dowshan Tepeh, east of Tehran, built in the opening years of this century

51. Greenhouse of the Dowshan Tepeh Palace

52. *The Abode of Pleasure, with Its Sleeping Palace and Separate Pavilions for Royal Favorites*, painting, 1880 (photo: ASI)

53. *The Bagh-i-Shah, Erected at Tehran in the Last Quarter of the 19th Century*, painting (photo: ASI)

54. *Terrace and Royal Living Quarters of the Castle of the Qajars*, engraving, Coste, pl. LVIII

55. *Central Kiosk in the Main Garden Area of the Castle of the Qajars*, lithograph, Flandin, pl. XXVII

56. *Palace Pavilion of the Royal Garden at Shahrestanak*, painting, ca. 1880 (photo: ASI)

57. A section of the modern royal gardens at Shimeran

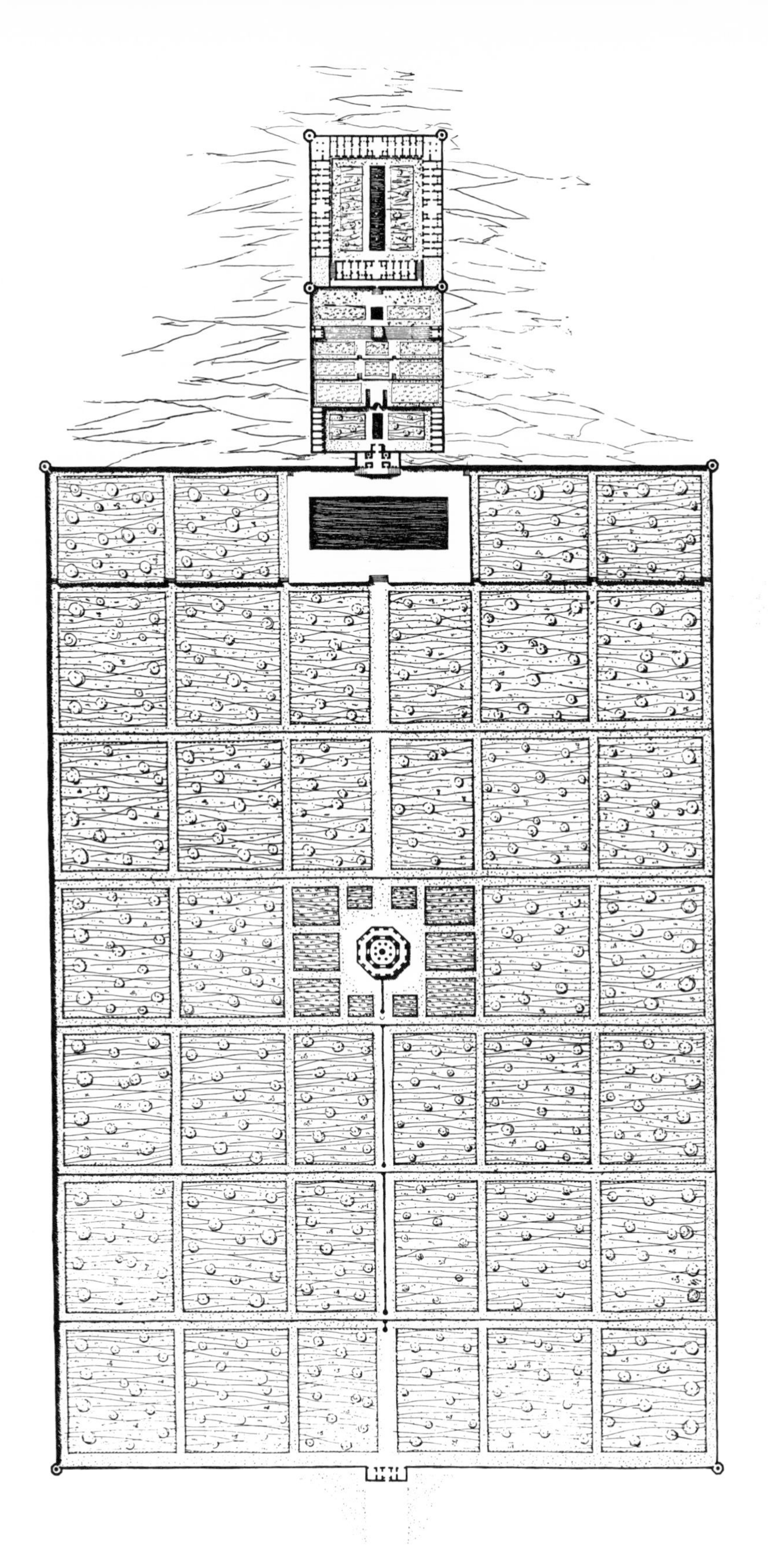

58. Restored plan of the Castle of the Qajars, northeast of Tehran

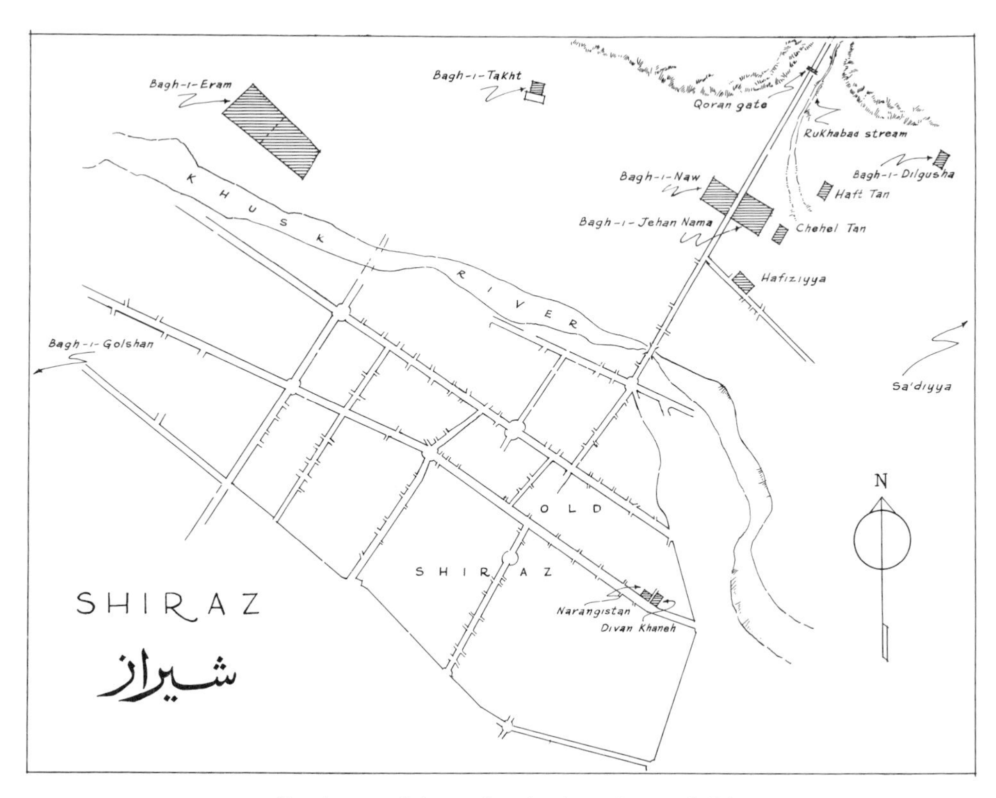


59. Sketch map of the garden sites in and around Shiraz

60. The old Qoran Gate at the pass overlooking Shiraz, now replaced by a modern archway

61. Corner of the pavilion in the Bagh-i-Dilgusha, a garden on the outskirts of Shiraz

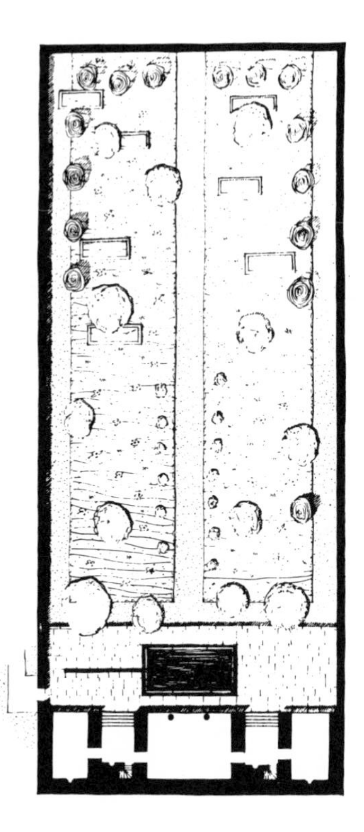

62. Plan of the garden and pavilion of the Haft Tan at Shiraz

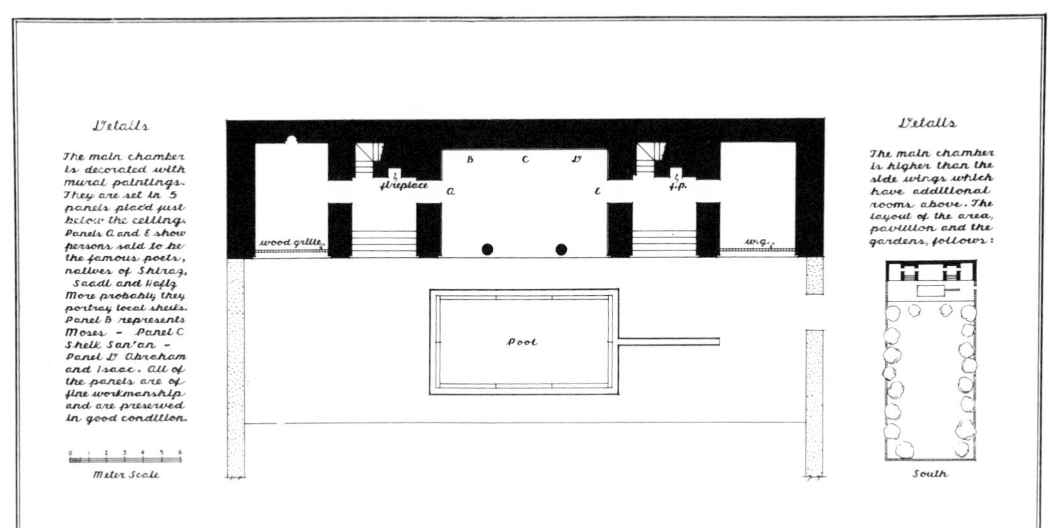


63. Detailed plan of the pavilion of the Haft Tan

64. Detail of stone carving at the site of the Bagh-i-Naw; the romantic lover Manjun living among the birds and animals

65. *The Bagh-i-Naw at Shiraz*, lithograph, Flandin, pl. LXXXV

66. The Bagh-i-Takht at Shiraz at about the turn of the century

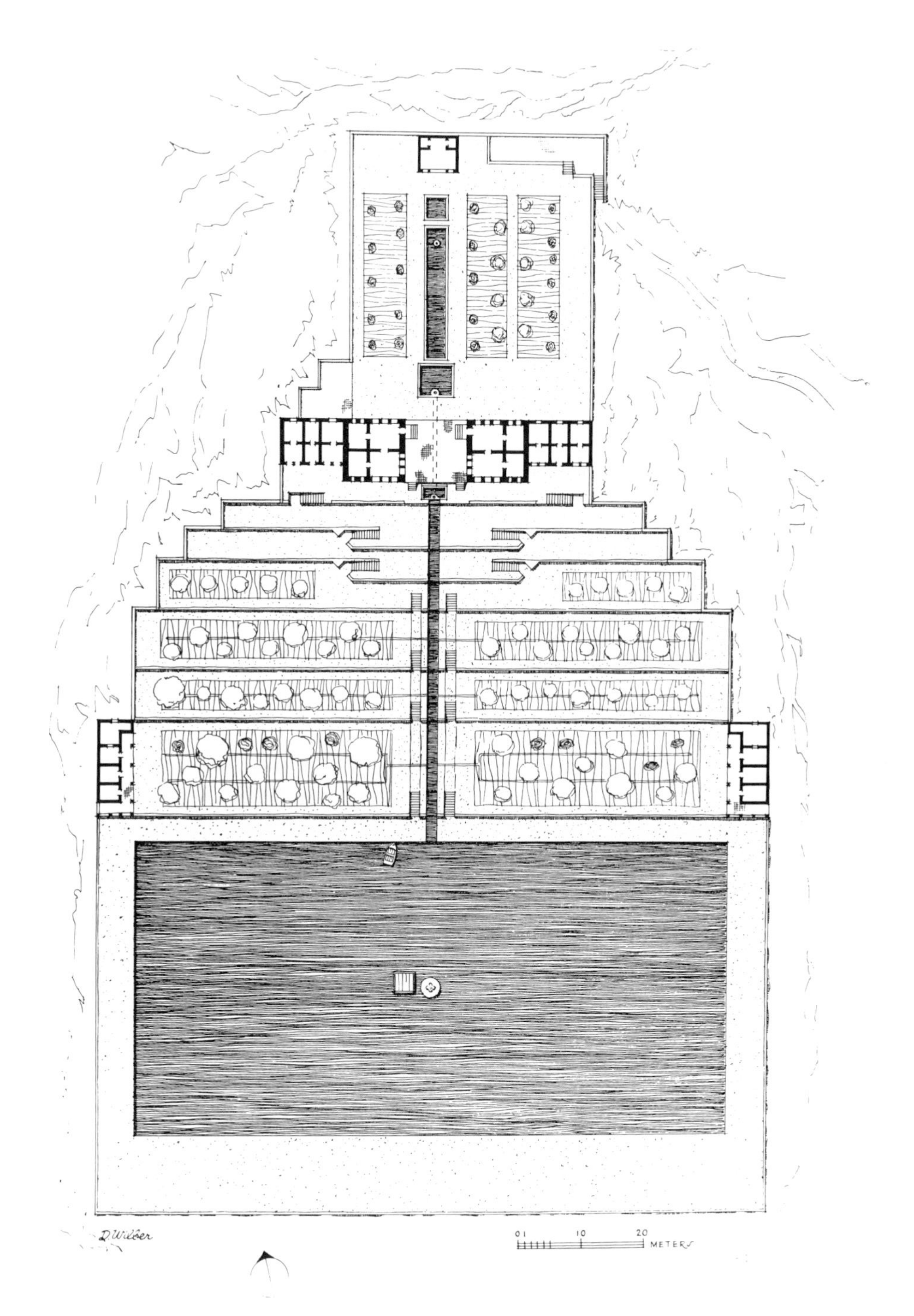


67. Restored plan of the Bagh-i-Takht

68. The main pavilion of the Bagh-i-Eram at Shiraz, shown with a former owner in the early 20th century

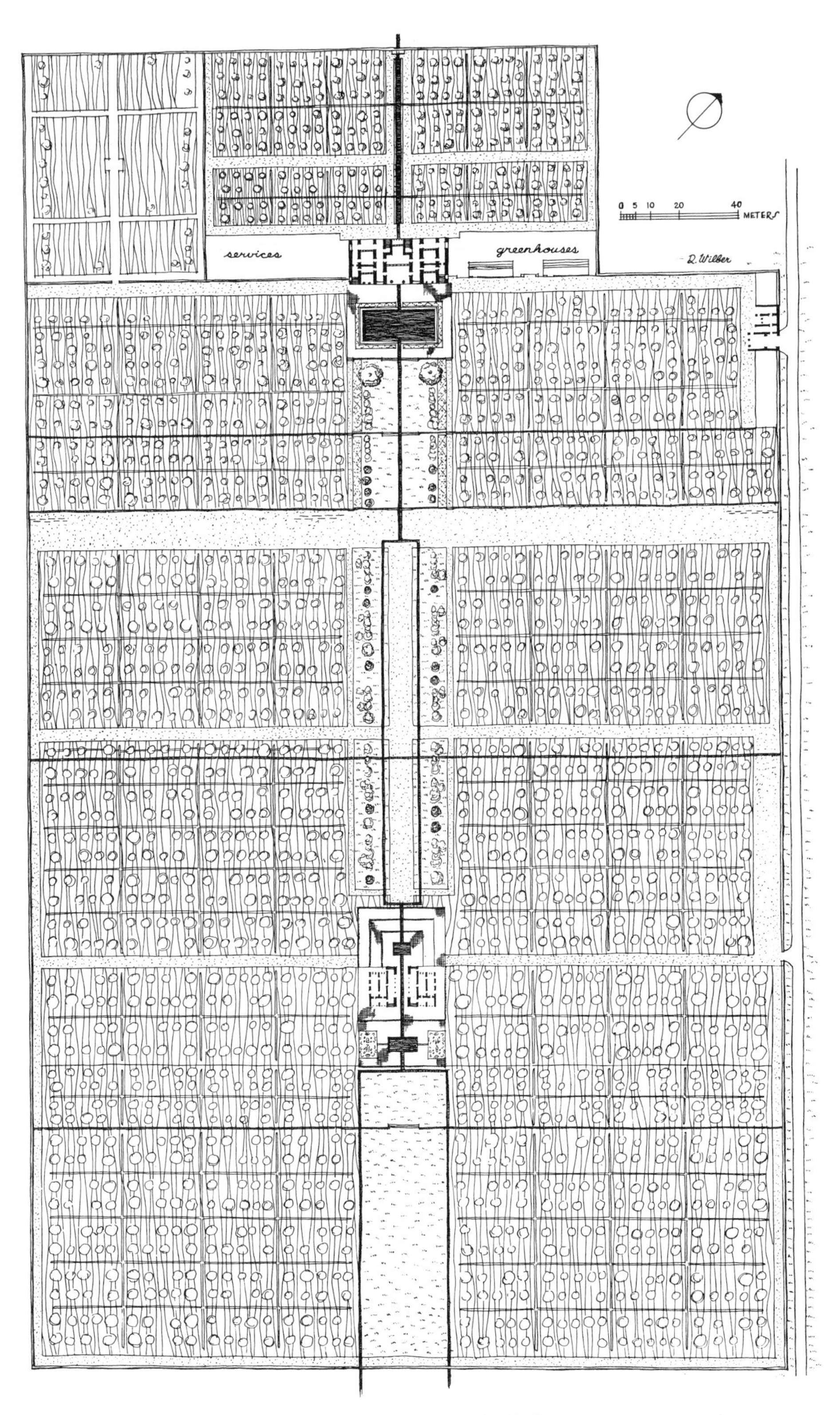


69. Restored plan of the Bagh-i-Eram; the area is divided into two separate enclosures by a wall at right angles to the long axis

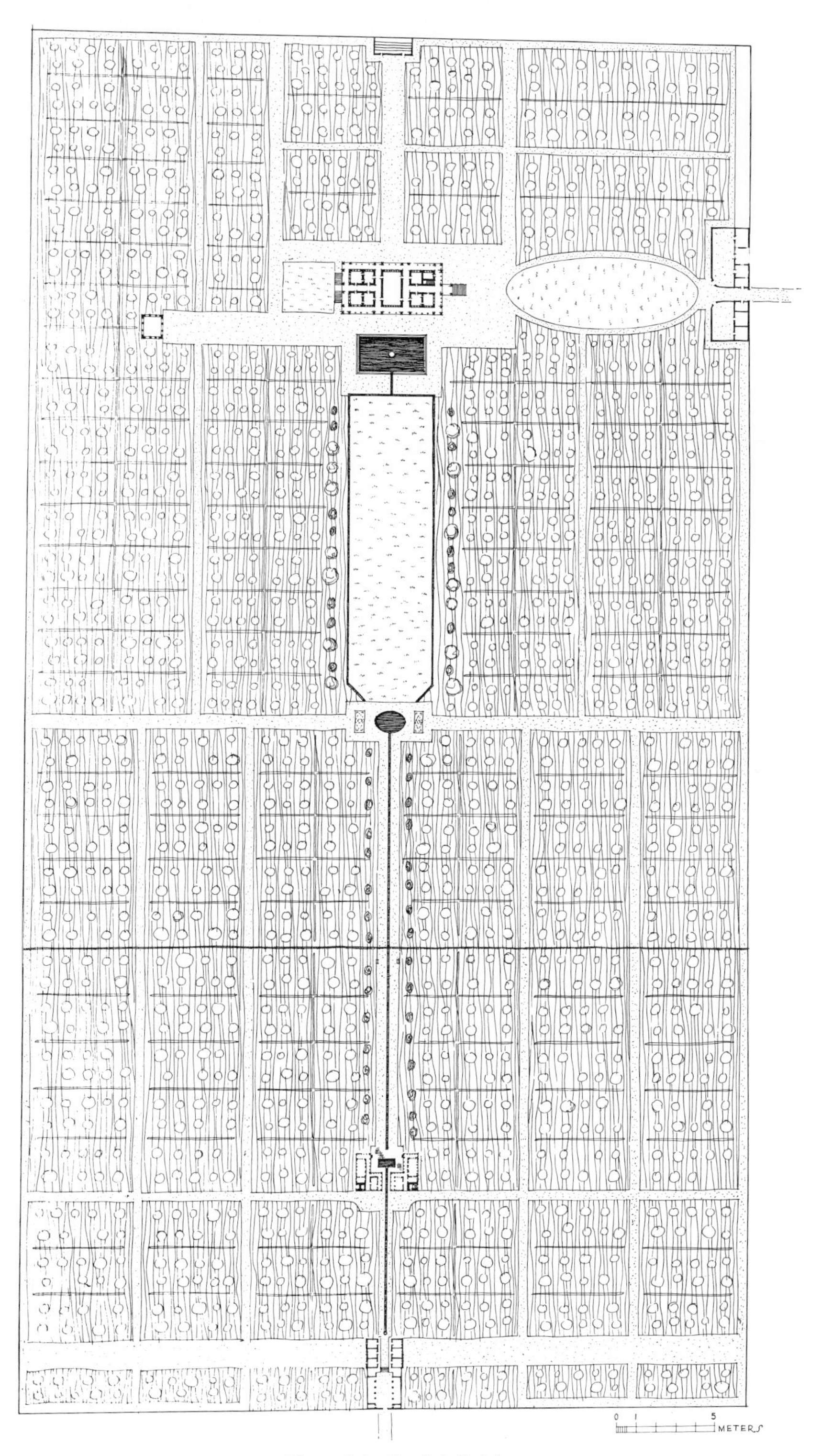


70. Plan of the Bagh-i-Gulshan

71. Main entrance to the principal garden pavilion in the Bagh-i-Gulshan

72. Now-vanished garden palace at Shiraz

73. Plan of the Divan Khaneh

74. Enameled tiles in the Divan Khaneh

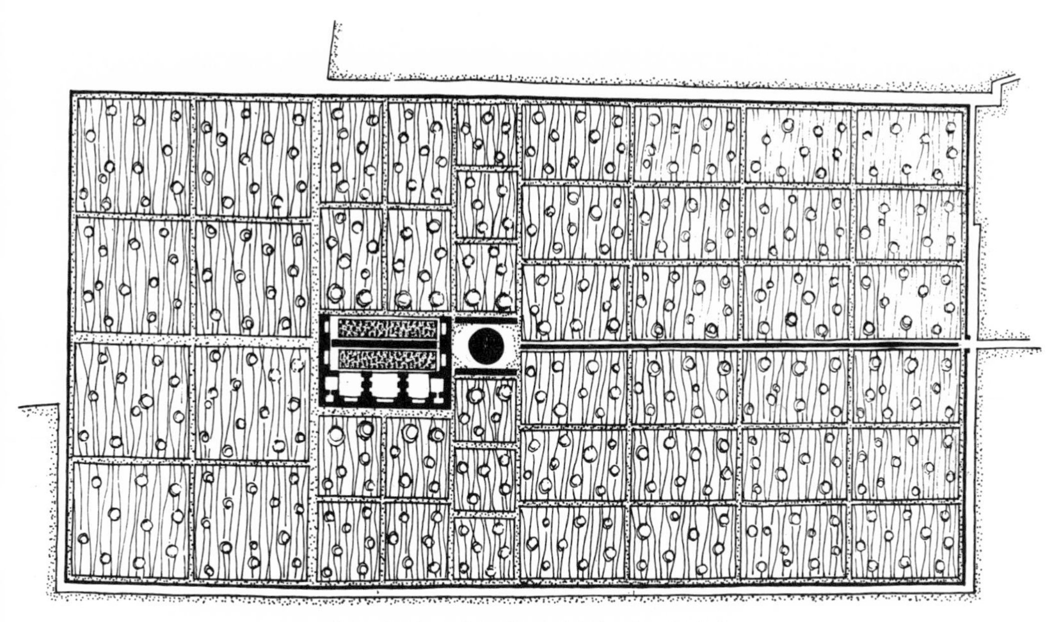

75. Sketch plan of the Bagh-i-Shimal at Tabriz

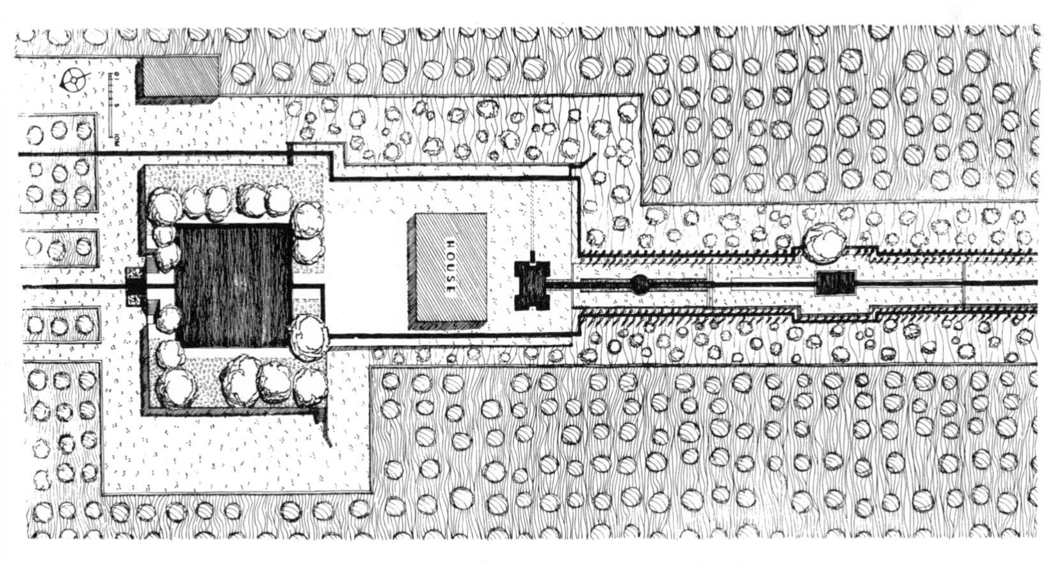


76. Plan of the garden at Fathabad, near Tabriz

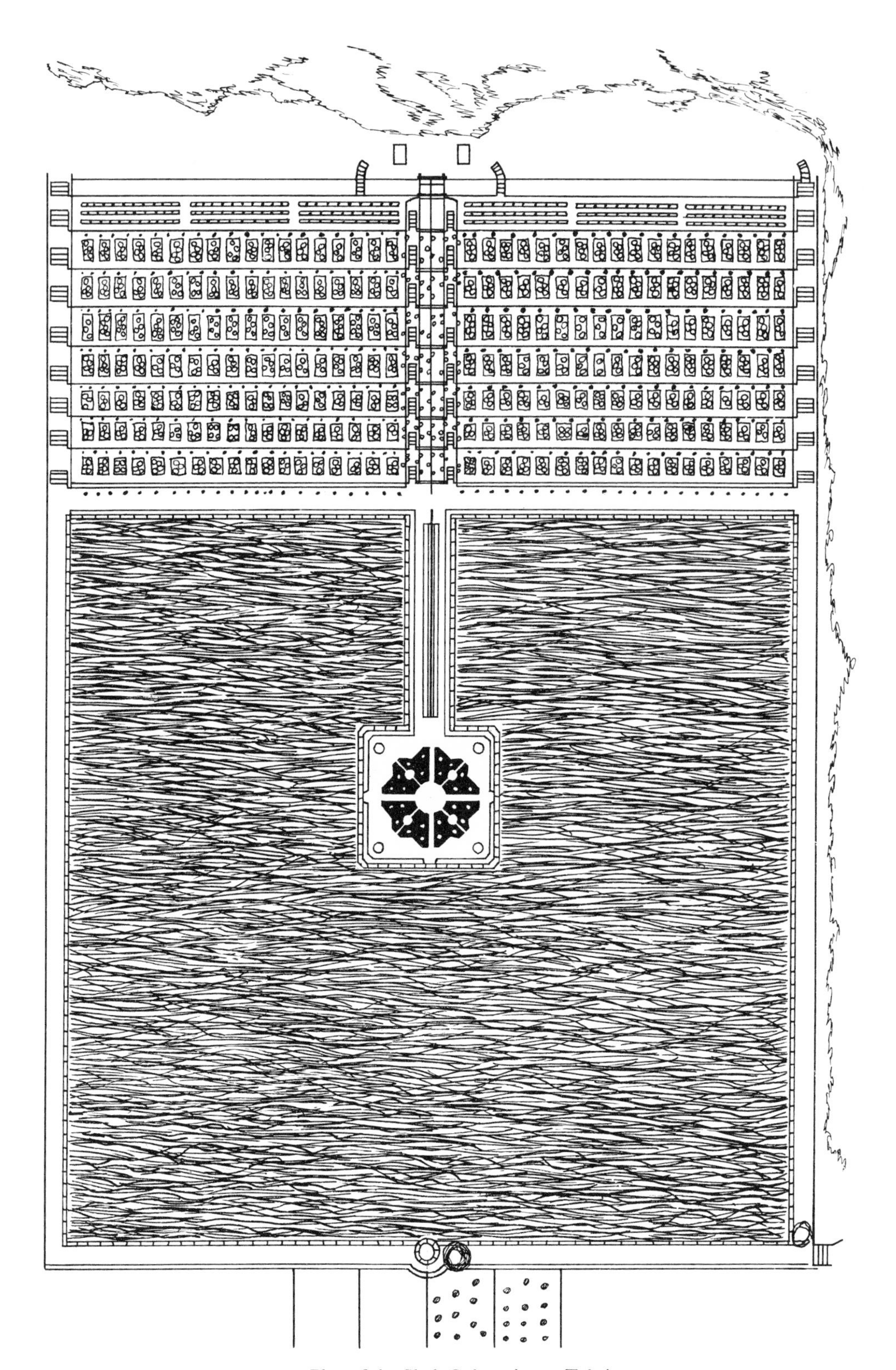

77. Plan of the Shah Gol garden at Tabriz

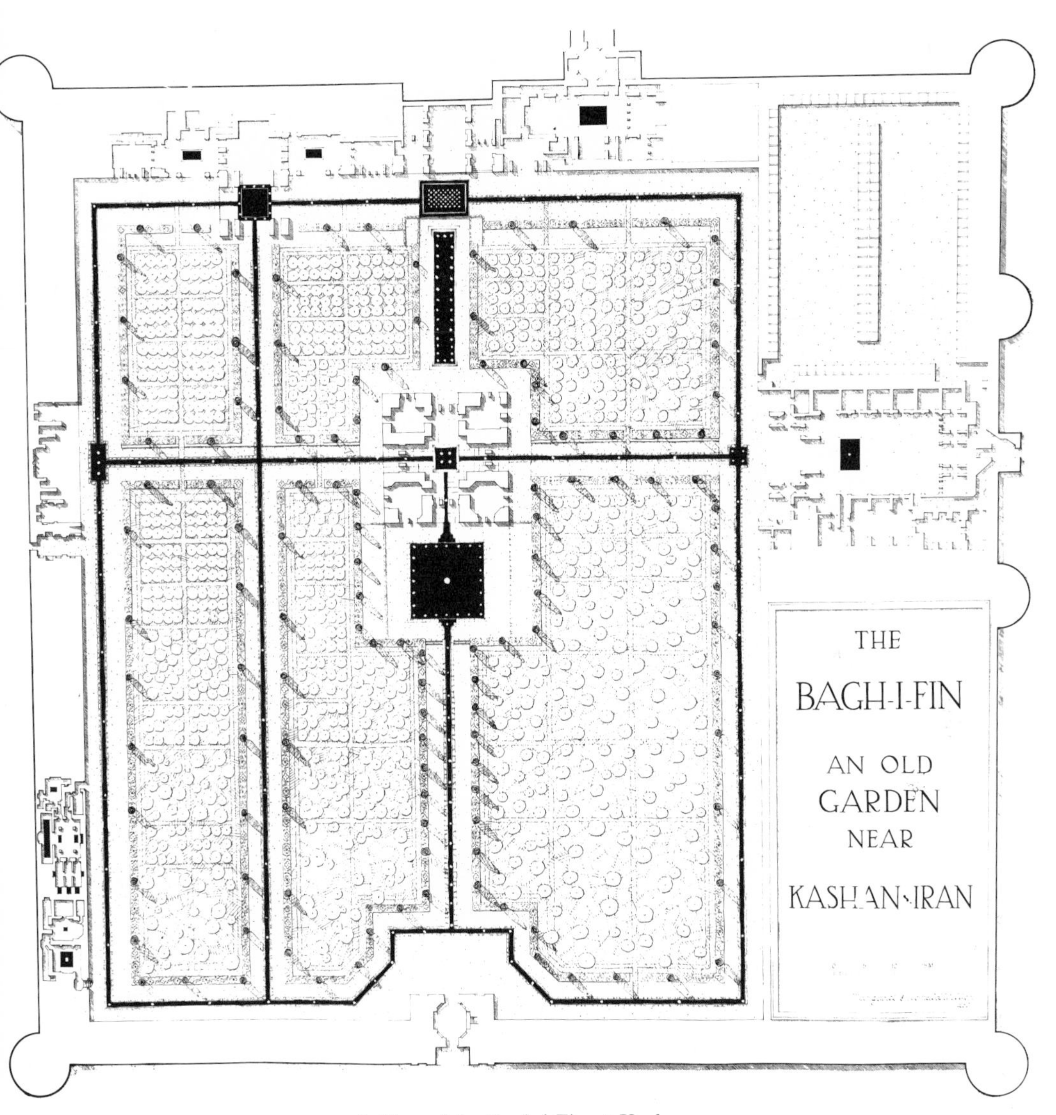


78. Plan of the Bagh-i-Fin at Kashan

79. The pool, fed by a qanat, which supplies water to the Bagh-i-Fin

80. The largest of the pools in the Bagh-i-Fin

81. *The Main Pavilion of the Bagh-i-Fin*, engraving, Coste, text illus. p. 38

82. A section of the Bagh-i-Fin showing walks, water channels, and bordering plane and cypress trees

83. View of the shrine at Mahan from one of its minarets

84. Typical pool in one of the courts of the shrine at Mahan

85. The oasis garden at Tabas, where the water gushes forth in the heart of the vast salt desert

# Index

*Where Persian and English names for a single garden have been used alternately in the text, all page references appear in the index under the Persian name with its English equivalent following in parentheses. Otherwise, gardens will be found listed under either the Persian or English name, depending on which appears exclusively in the text.*